WARM-UP PREVIEW

THE PROCESS OF DESIGN

Some people call engineers "problem solvers extraordinary." While this is an exaggeration, engineers do have a knack for solving problems. They acquire this knack through years of practice—starting with four or five years in college. It is there that they learn that problem solving (which is the very heart of engineering and which, along with modeling and optimization, is the subject of this unit and the next) is a skill.

What you will be introduced to in this unit is the five-phase problem-solving procedure that engineers generally refer to as the *design process*, or simply *design*. In sum it consists of:

- *Problem formulation*. Broad, detail-free definition of the problem at hand.
- *Problem analysis*. Detailed definition of the problem.
- *Search for solutions*. Accumulation of alternative solutions through invention, inquiry, and the like.
- *Decision*. Evaluation, comparison, and screening of the alternatives until the best one evolves.
- *Specification*. Complete documentation of the chosen solution.

Learning Segment 3
FORMULATING AND ANALYZING THE PROBLEM

LEARNING OBJECTIVES

- Understand the meaning of the following terms: *state A, state B, problem, solution, formulation breadth, solution restrictions, solution criteria, input, output black box, variables*, and *constraints.*

- Be able to formulate a problem as it is done in the design process.

- Be able to perform a problem analysis as it is done in the design process.

WHAT CONSTITUTES A PROBLEM?

Do you advocate trying to solve a problem without knowing what the problem is? Surely you say no. Yet many people do exactly that in practice. In casual problem solving this habit matters little but, when problem solving is your business, it can be consequential, indeed. Thus it is important that an engineer clearly define a problem before becoming involved with solutions.

To do so, however, requires specific notion of what constitutes a problem. Many people have never given that word much thought, and why should they? In conversational use a strict definition is hardly necessary; everyone knows more or less what the word means, and that suffices. But it does not suffice for an engineer attempting to characterize a specific problem prior to solving it. Then a fuzzy idea of what constitutes a problem is a handicap.

A problem arises from a desire to achieve a transformation from one state of affairs to another. These states might be two locations, the interval between which must be traversed (two points on opposite sides of a river, one city and another, Earth and another planet). Then, too, many problems involve a transformation from one form or condition to another (for example, brakish to potable water). In every case there is an originating state of affairs (call it "state A") and a state of affairs the problem solver seeks a means of achieving (call it "state B"). Figure 1 provides examples.

FIGURE IT OUT

Study the examples of state A and state B in Figure 1. Use the same sort of logic to fill in

MODERN ENGINEERING
Design: Skills in Problem Solving

Edward V. Krick

Lafayette College
Easton, Pennsylvania

Wiley Professional Development Programs

Advisory Editor
Steven C. Wheelwright
Harvard Business School

John Wiley & Sons, Inc.
New York • London • Sydney • Toronto

Library of Congress Catalogue Card Number: 76-10038

ISBN 0-471-01702-7

Printed in the United States of America.

10 9 8 7 6 5 4 3 2 1

the missing examples of state A or state B in the chart below. You will recall that these examples were discussed in the first unit of this program.

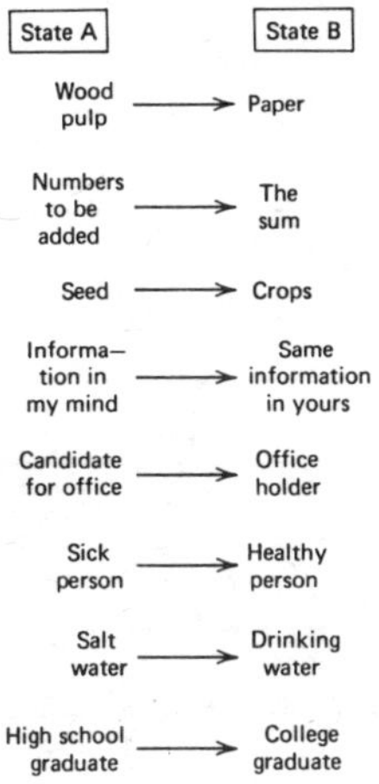

State A	State B
Wood pulp	Paper
Numbers to be added	The sum
Seed	Crops
Information in my mind	Same information in yours
Candidate for office	Office holder
Sick person	Healthy person
Salt water	Drinking water
High school graduate	College graduate

Figure 1

State A	State B
______________________ ______________________ ______________________	1. You, a slim, trim, muscular type, at an ideal weight for your age and height.
______________________ ______________________ ______________________	2. Heat energy.
______________________ ______________________ ______________________	3. You, fully equipped with the engineering skills and knowledge described in this course.
Enough glass tubing, reed conductors, and inert gas for millions of switches	4.______________________ ______________________ ______________________

1. Only you can answer this, but, if you are like most of us, you will have mentioned excess and ill-distributed weight.

2. You could mention oil, coal, gas, sun, or some other familiar source, but we would be happier if you simply wrote latent source of heat.

3. Only you can characterize yourself in these respects.

4. Millions of the specified reed switches satisfactorily assembled.

WHAT CONSTITUTES A SOLUTION?

A *solution* is a means of achieving the desired transformation. It is difficult to envision a problem to which there is only one solution; for most problems there are more alternative solutions than there is time to investigate. Think of the many modes of travel and all the possible routes they can be combined with to provide alternative means of getting from one point on the map to another.

TECHNICAL TERMS

Write a brief definition for each term.

1. State A ______________________________

____

____

2. State B ______________________________

____

____

3. Problem ______________________________

____

____

4. Solution ______________________________

____

____

1. State A is the state or condition that is to be altered in some way.

2. State B is the state or condition into which state A is to be transformed.

3. A problem arises from the desire to achieve a transformation from one state (state A) to another (state B). (Of course, it takes more than a desired transformation to make a problem. For one thing, there must be alternative means of achieving the desired transformation. There must also be a preference between these alternatives. Without such a preference, there is no problem, for any one of the solutions will do.)

4. Solution refers to a means of achieving the desired transformation.

Criteria for Solutions

But that's not all; a problem involves more than finding *a* solution; it requires finding a preferred means of achieving the desired transformation (the mode of transportation that is best with respect to cost, speed, safety, comfort, and reliability). A basis of preference among solutions is commonly referred to as a *criterion*. Some criteria employed in the VTOL problem are the plane's cost, safety, reliability, noise level, maneuverability, and speed. Engineers are accustomed to thinking about, evaluating, and struggling with criteria.

Restrictions on Solutions

Engineers are also accustomed to living with another common characteristic of problems: *restrictions*. A restriction is an inescapable solution characteristic—a "must" for a solution if it is to be eligible for consideration. The first unit provides examples, like "must produce results within one minute." You can spot others in the case histories you will find throughout the course.

FIGURE IT OUT

Complete the following chart so that for each person and problem there is a state A, a state B, a criterion for the solution, and a restriction on the solution.

PERSON AND PROBLEM	STATE A	STATE B	CRITERION	RESTRICTION
1. Business person's problem	An invested sum of money			
2. Missionary's problem	A tribe of cannibals			
3. Defense lawyer's problem	Twelve undecided jurors and a guilty client			
4. Newspaper deliverer's problem				
5. TV repairer's problem: a house call on the day of the big game				

1. The business person's state A is an invested sum of money; state B will be the return of the money with earnings; a criterion would be to maximize the earnings; a restriction might be not to exceed a certain degree of risk with the money.

2. The missionary's state A is a tribe of cannibals; state B might well be a tribe of natives converted to the missionary's religion; a criterion for the solution might be the period of

time required; one restriction on the solution would certainly be that the conversion take place without the missionary being eaten.

3. The defense lawyer's state A is the twelve undecided jurors and his or her guilty client; state B would be a unanimous vote of acquittal by the jury members; one criterion of the solution might well be the probability of success (certainly not all defense schemes are equal in this respect and certainly this probability matters); one restriction on the solution might well be that the defendant not be allowed to testify in his or her own defense, lest cross-examination force an unintended confession of guilt.

4. The newspaper deliverer's state A might well be a bundle of undelivered newspapers; state B would probably be newspapers delivered safely to the homes of the subscribers; one criterion might be the amount of time required; one restriction might be that none of the papers be deposited on a subscriber's roof.

5. The TV repairer's state A is a malfunctioning set in a customer's home; state B is a properly functioning set in the customer's home; one criterion of the solution might be cost; one restriction might well be that the set not be taken to the shop, because the customer wants it working in the home before the big game begins.

There are still other problem characteristics you must cope with, but by now it should be apparent that as a professional problem solver you must have specific characteristics in mind when thinking about a problem. It follows, too, that you had better have an effective method of solving problems. To help you develop one, here is a short discourse on problem solving.

PROBLEM SOLVING

First Define the Problem—Don't Jump to Solutions

The management of a company that distributes livestock feed is concerned about the high cost of handling and storing its products. At present the materials are bagged and stored by the procedure diagrammed in Figure 2. An engineer is assigned to find a less costly method.

A common tendency is immediately to begin thinking of improvements in the present solution. In this problem, you must admit, it is tempting to begin by scrutinizing the solution described in Figure 2, seeking improvements that might reduce the cost of the process. If you were to do so, you would become concerned with matters such as equipment for filling, weighing, and sewing the sacks, arrangements of facilities, ways of transporting the heavy sacks, means of combining operations, and other possible improvements. This is exactly what not to do—immediately becoming entangled in details of possible solutions. In so doing you would be generating solutions to a problem you have never defined.

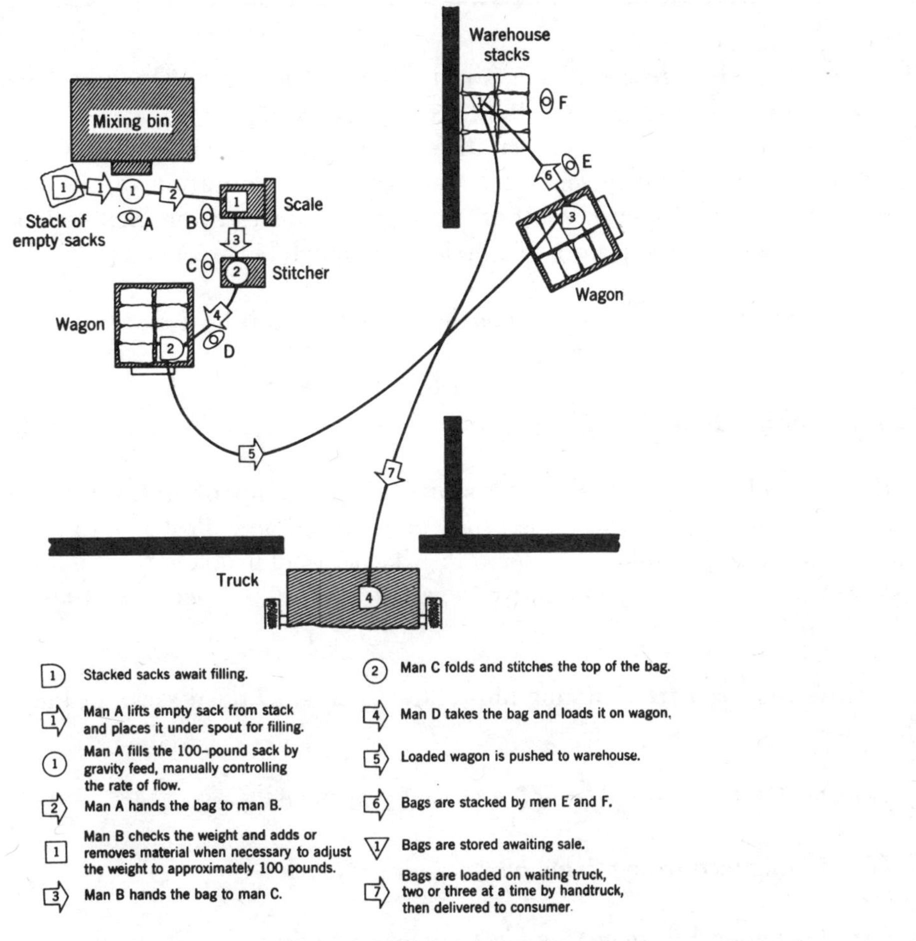

Figure 2. The present method of filling, storing, and loading sacks of feed.

THINK IT THROUGH

It is important, first, to define the problem. Getting entangled in the current solution to a problem prevents one from defining the problem. Why is this so?

____

____

____

____

____

An answer to this question appears in the following text.

The current solution to a problem is not the problem itself. This may seem too obvious to say, yet many people apparently don't see the distinction. They attack the present solution, not the problem. There is a crucial difference between picking at a current solution in an effort to eliminate inadequacies and starting with a definition of the problem and from there synthesizing a superior solution through an intelligent problem-solving procedure. In the long run the latter approach will make for a much better engineer.

So now you know how not to go about it. How should it be done?

Problem Formulation

At the start, identify the desired transformation in terms of states A and B. This first step in the problem-solving process is *problem formulation*. Problem formulation should be made as broad as possible. To illustrate what a broad problem formulation is and why it is important, consider these alternative views of the feed problem: To find the most economical method of

1. Transferring feed from mixing bin (state A) to stockpiled sacks in the warehouse (state B).

2. Transferring feed from *mixing bin* to sacks on the delivery truck.

3. Transferring feed from *mixing bin* to delivery truck.

4. Transferring feed from *mixing bin* to delivery medium.

5. Transferring feed from *mixing bin* to the consumers' storage bins.

6. Transferring feed from storage bins of the feed ingredients to the *consumers' storage bins*.

7. Transferring feed from producer to consumer.

THINK IT THROUGH

Broadening the formulation of a problem tends to remove restrictions from its possible solutions. Can you see any differences in the solutions that might result from the seven formulations of the problem given above?

The answer to this question appears in the following text.

These formulations of the problem are not equally advisable; the probable consequences of pursuing each are quite different. In formulations 1 and 2, state B is *feed in sacks*. In transformation 3 only "truck" is specified, which opens the problems to solutions not involving sacks. In formulation 4 only "delivery medium" is assumed to state B, opening up additional possibilities. This trend toward a less specific definition of states A and B continues until only producer and consumer are specified, leaving the way open for a wide variety of methods of handling, modes of transportation, package types, and so on. From all this it is apparent that as the specifications assumed for states A and B become more general, the alternative solutions become more numerous and varied. Narrow problem formulations cause whole realms of profitable possibilities to be unnecessarily excluded from consideration. Most persons trying to solve the feed problem would automatically and unjustifiably assume state B to be stockpiled sacks in the warehouse; they would solve the problem without realizing that they themselves have limited the problem to this extent.

In formulation 1, state B goes only as far as the warehouse stockpile. Formulation 2 extends state B to the truck, and formulation 5 extends it to the consumer. In formulations 6 and 7 state A is extended. In these instances the problem is extended to embrace more of the total problem. In general, strive to formulate each problem *to include as much of the total problem as the importance of the situation and organizational boundaries will permit.* This is for a good reason. The more the total problem is split into subproblems to be solved separately, the less effective the total solution is likely to be. If bagging the feed is treated as one problem, transportation to and stacking in the warehouse as another, transportation to the consumers as another, and unloading of trucks as still another, the overall feed distribution system that eventually evolves will probably be far from optimum. Treating this problem broadly is very likely to result in a superior solution overall.

The detail in which states A and B are specified and the proportion of the total problem they encompass will hence forth be referred to as the *breadth* of the problem formulation. In the case of this problem, formulation 7 is far broader than formulation 1.

TECHNICAL TERMS

Write a brief definition for each term.

1. Breadth of problem formulation ______

2. Restrictions on solutions ______________________________

____

____

3. Criteria for solutions ______________________________

____

____

____

1. The breadth of a problem formulation is the degree of detail to which states A and B are specified and the proportion of the total problem encompassed by the formulation.

2. Restrictions on solutions are conditions that must be met by the solution. The broader the formulation of the problem, the less restrictions, generally, it will impose on the possible solutions.

3. Criteria for solutions are standards by which the success of the solution will be measured.

Importance of a broad formulation. The engineer assigned to this project succeeded in freeing himself from the limitation of using sacks and, thereby, he opened the problem to the possibility of handling feed in bulk. (Now you may think that's no feat. If so, you grossly underestimate the thought-restricting role of customary—especially longstanding—solutions to problems.) Also, his formulation encompassed delivery to the consumer, which opened the way to delivering feed in bulk directly into the farmer's storage bins. The result is that after many years of back-breaking loading and unloading of heavy sacks, dealers now deliver feed by blowing it through a tube leading from a bulk-delivery truck, a large "feed bin on wheels," directly into the farmer's storage bins.

Broad treatment of problems that previously were attacked in piecemeal fashion can pay off handsomely. Figure 3 provides an indication of what can come of the broadening of some familiar problems. We are surrounded by problems that are unsatisfactorily solved, mainly because the solvers are traditionally "narrow-sighted." This applies to education, business, medicine, and every other field, as well as to engineering. The reason we are making this plea for broad problem formulations is that the probability of vastly improved solutions is high. Society is in need of engineers who will attack problems in an unconventionally broad manner.

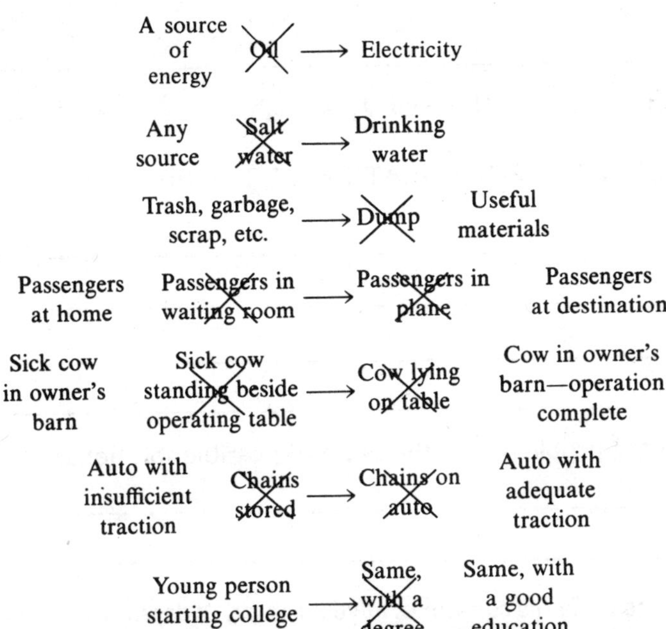

Figure 3. Some formulations of familiar problems, indicating how each might be broadened. Perhaps you can visualize how these broadenings can pay off.

How broadly you formulate a given problem is your decision. You can choose a very broad view that maximizes the number and scope of alternative solutions that can be considered. Or you can go to the other extreme and choose a formulation offering very little latitude in the way of solutions. Remember, however, a problem formulation is a point of view—the manner in which you perceive a problem. It may be no more than some thoughts or some scribbled notes. It is not irrevocable; it can be changed if you later find it necessary or desirable. Therefore you should formulate problems broadly; it is your perogative—in fact, your professional obligation—to do so. You are selling yourself and your employer short if you don't.

CHECK IT OUT

Consider a problem that your company has had to deal with recently. Try to recall, as exactly as you can , how the problem was stated. Then try to improve the statement of the problem as indicated by the questions below.

1. PROBLEM AS ORIGINALLY FORMULATED

State A ______________________________

____

Was state A formulated too broadly? ____ Not broadly enough? ____ Okay? ____

State B ______________________________

Was state B formulated too broadly? ____ Not broadly enough? ____ Okay? ____

2. A POSSIBLE IMPROVEMENT IN THE FORMULATION OF THE PROBLEM

State A ____

____

State B ____

____

How would this reformulation of the problem alter the range of possible solutions?

____

____

There is, of course, no way for us to verify the correctness of your answer. It depends on how accurately you have recalled your company's formulation of the problem in terms of state A and state B, and how well you have applied the concepts presented in this section in your reformulation of the problem.

If someone else from your company is taking this course, it will be very helpful for you to discuss your answers with him or her. Even if there is no one else, you may be able to discuss the concepts in this exercise with a fellow worker. A good time for such a discussion might be over lunch or during breaks.

Methods of formulating a problem. A problem can be formulated verbally or diagrammatically, on paper or in your mind. In many instances a few words will suffice. Or maybe you prefer a simple diagram. The black-box method of viewing a problem is a diagrammatic formulation. The usefulness of this approach can be illustrated by applying it to a type of problem that is often unsatisfactorily defined: an information-processing problem. An office that handles airplane reservations is an information-processing system. A customer comes to a ticket agent with a request in terms of number of seats, date, and perhaps other specifications. These constitute the input to the black box—state A, in other words. The output—state B—is also information, in the form of confirmation of the request or a quotation of the available alternatives. In the problem-formulation stage, what transpires within the black box is not known or of interest. The box replaces the details you are trying to avoid at this stage, and that is the key to its helpfulness.

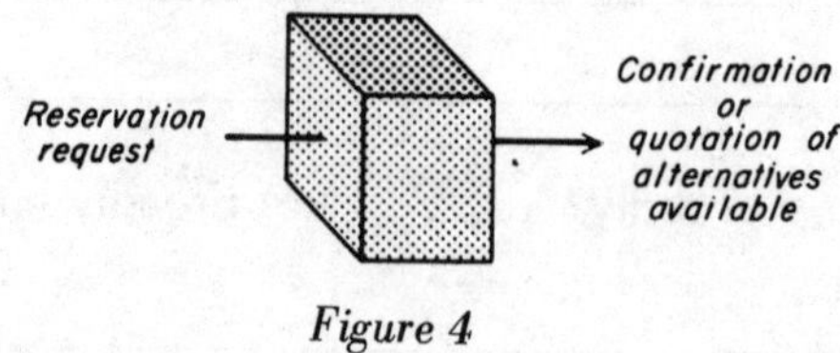

Figure 4

About now it must occur to you that there are no hard and fast rules for problem formulation. There is no such thing as *the* formulation of a given problem, but there certainly are more and less profitable formulations. The best that can be done is to offer you guidelines and examples. It is up to you to benefit from these and from your own experience in developing your problem-formulating skills.

APPLY YOUR SKILLS

For each solution given below provide the input (that is, state A) and the output (state B) of the problem. This exercise will give you some practice formulating problems in terms of state A and state B.

STATE A—INPUT	SOLUTION	STATE B—OUTPUT
1. ________	Gearbox	________
________		________
________		________
2. ________	Electric iron	________
________		________
________		________
3. ________	Air conditioner	________
________		________
________		________

STATE A—INPUT	SOLUTION	STATE B—OUTPUT
1. Shaft rotating at one speed	Gearbox	Shaft rotating at another speed
2. Damp and wrinkled clothes	Electric iron	Dry and smooth clothes
3. Warm and humid air	air conditioner	Cool and dry air

TECHNICAL TERMS

Write a brief definition for each term.

1. Input ________________________________

2. Output ____

____

3. Black box ____

____

1. Input, in problem solving, means the same thing as state A; it is the state of affairs in which a problem arises for the problem solver. Input is the information that is fed into a black box for solution.

2. Output, in problem solving, is the same thing as state B; it is the state of affairs that the problem solver seeks a means to achieve. Output is the result of the solution that is to flow from the black box.

3. Black box, in problem solving, means a diagrammatic element that takes the place of all the details of the problem solution(s) and that is placed between the input and output elements in the diagrammatic approach to problem formulation.

From now on, when you are approaching a problem, hopefully you will, at the outset, carefully choose the problem you intend to solve and, in so doing, give thoughtful consideration to alternative formulations and the probable consequences–payoffs as well as obstacles–of each. Strive for a broad, uncluttered view of a problem before you concern yourself with possible solutions, in fact, before you become entangled in *any* details. After you have done so the problem can be–must be–defined in detail.

Problem Analysis

The following case illustrates defining a problem in detail. An appliance manufacturer has tentatively decided to market a new type of clothes washer. This machine is supposed to perform the usual tasks expected of such machines and also to serve as a home dry-cleaning unit. Management has also decided that:

1. This unit must not be larger than 80 centimeters wide, 80 centimeters deep, or 100 centimeters high.

2. It *must* operate on 60-cycle, 115-volt alternating current.

3. It *must* be approved by Underwriters Laboratories.

4. The cost of manufacture *must not* exceed $125.

5. It *must* satisfactorily process all natural and synthetic textiles.

6. It *must* be foolproof in operation.

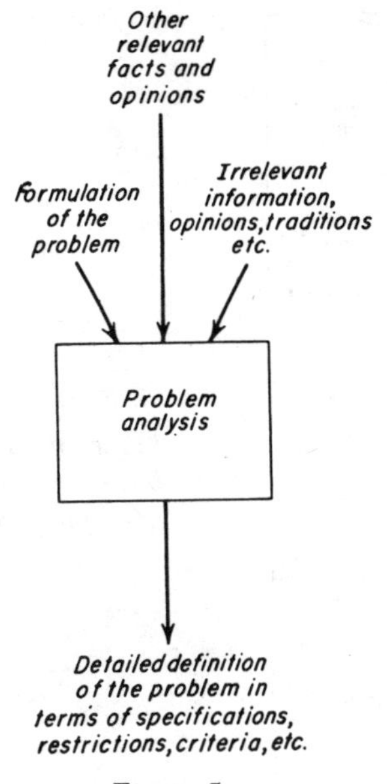

Figure 5

MEMORY JOGGER

We have introduced a technical term to refer to the kind of thing that is represented by each of the six items decided upon by management. What is that term and its definition?

It is a restriction on a solution, which is a condition that the solution must meet to be at all acceptable.

In formulating this problem, the engineer assigned to design this multipurpose cleaning machine identified state A simply as dirty fabrics and state B as clean fabrics. However, to solve this problem, he had to learn a lot more about it. He did this through considerable deliberation, investigation, and consultation, especially with marketing experts, who are in close touch with consumer preferences. The result of these efforts is the problem analysis shown in Figure 6. Here is an interpretation of his notes.

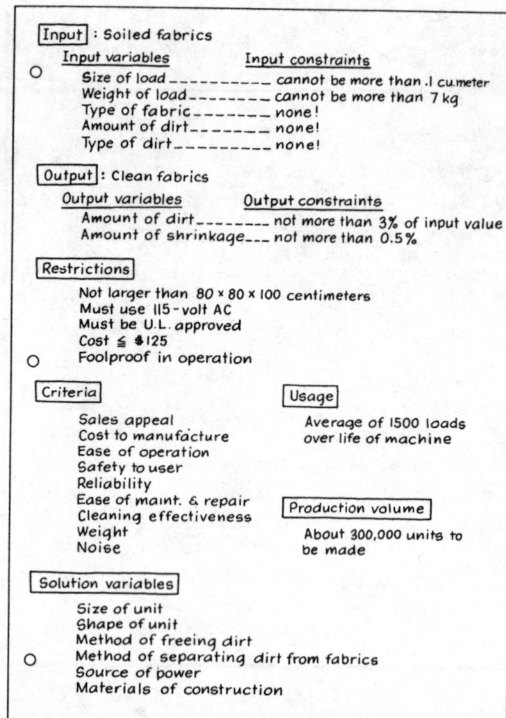

Figure 6. This is a page from the engineer's notebook, summarizing his analysis of the cleaning machine problem.

Input. An important part of the engineer's problem analysis is getting detailed information on the input and output (state A and state B). Here is an example of the detail required: Input and output tend to vary over time, which is certainly true for a washer. The amount of clothing a person places in the machine varies from load to load, and so do the types of fabrics and amount of dirt. This problem is typical in this respect; very few input and output characteristics remain constant in any problem. (Examples from other problems: the ore input to a steel mill varies in chemical makeup; the electrical output of a power plant certainly varies over a period of time). These dynamic characteristics of states A and B are called *input variables* and *output variables*.

Often there is a limit on the extent to which an input or an output can fluctuate. In this problem an upper limit of 7 kilograms has been set for the input variable "weight of load." This is called an *input constraint*; the equivalent for state B is an *output constraint*. Before an engineer can solve a problem satisfactorily, he must have reliable estimates of input and output variables and their constraints.

Restrictions. A restriction is a solution characteristic previously fixed by decision, nature, law, or any other source the problem solver must honor. Nature dictates that light, water, and nutrients must be provided to transform seeds into plants. Through building codes, certain characteristics of structures are fixed by law. Some restrictions limit the problem solver's choice to a *range* of values: for example, "the machine cannot be greater than 80 × 80 × 100 centimeters." Others *fix* a solution characteristic: "it must operate on 60 cycle, 115 volt A.C." Therefore, solutions larger than 80 × 80 × 100 centimeters are ineligible, as are solutions using other than the specified power source.

TECHNICAL TERMS

Write a brief definition of each term.

1. Variables ____

2. Constraints ______________________________________

1. Variables are allowable changes in input or output.
2. Constraints are the limits within which variables in input and output must remain.

Criteria. The criteria to be used in selecting the best solution should be identified during problem analysis. Actually, criteria do not change radically from one engineering problem to another; construction cost, safety, reliability, appearance, ease of use, and maintenance cost are criteria that apply in almost every case. But the relative weights of these criteria are likely to change significantly. Hence, in most cases, the engineer's main task is to learn the relative importance attached to various criteria by officials, customers, clients, citizens, and others. This information is important, as the following case will illustrate: Assume that safety is to be a heavily weighted criterion in the design of a rotary lawn mower. Knowing this, the designer will consider a different set of materials, mechanisms, cutter types, and discharge methods than he would if safety were not considered extremely important. An unusually weighted criterion, therefore, affects the types of solutions the engineer will emphasize in the next phase of the design process, the search for alternative solutions.

FIGURE IT OUT

You have previously written definitions for both criteria and restrictions. It is important to recognize how they differ for the problem solver. How do they differ?

An answer to this question appears in the following paragraph.

Some readers may find the distinction between restrictions and criteria fuzzy. But note that a *restriction* is an either-or proposition; a solution either satisfies a restriction or it doesn't. With a *criterion*, being a value by which solutions are judged, it is a matter of degree; some solutions are more or less attractive than others. A criterion lacks the clear-cut, go/no-go feature of a restriction. We look for the *least* costly, the *most* attractive, and the *safest* when cost, attractiveness, and safety are the criteria. In the design of an appliance, to say that energy consumption matters is to say energy consumption is a criterion. Other things being equal, the solution that consumes the least energy will be chosen. To say that any solution that consumes more than 10 watts is unacceptable sets up a restriction.

Solution variables. Alternative solutions to a problem differ in many respects. Solutions to the cleaning machine problem differ in size, shape, method of freeing dirt from fabrics, type of mechanism, materials of construction, and so forth. The ways in which solutions to a problem can differ are called *solution variables*. Strictly speaking, a solution variable is a solution characteristic that the problem solver is free to alter.

Usage. Still another type of information that the engineer must obtain or predict himself is the extent to which his solution will be employed. This is often referred to as *usage*, and it very definitely affects the type of solution that will prove to be best in the situation at hand. Usage becomes significant whenever total cost—the cost of arriving at a solution *plus* the cost of physically creating it *plus* the cost associated with using it—is a matter of concern; and when is this not the case? If a river is to be crossed only rarely at a given spot, a bridge is obviously not the solution that minimizes total cost. On the other hand, if over a period of time millions of persons will cross the river at this spot, a rowboat is hardly the preferred method with respect to the total cost criterion.

Production volume. Suppose that only 10 of these cleaning machines are to be built. Under these circumstances the designer would care little about the manufacturability of his creation. If nonstandard, expensive components were specified and hand methods of fabrication were required, this would be of little concern as long as the volume was only 10 machenes. However, 300,000 machines is a different story; under these circumstances the engineer will be vitally concerned with the manner in which alternative designs affect manufacturing cost. This number, referred to as *production volume*, has a significant effect on the type of solution that is best, and obviously the engineer should know what volume is expected before he starts thinking about solutions.

YOU DESIGN IT

Assume that you are designing a vegetable cannery. How do you formulate the problem? What information concerning input and output would you gather during your analysis of the problem? What are some of the major criteria that you believe should be employed?

1. Input information ______________________________

2. Output information ______________________________

3. Some criteria ______________________________

TECHNICAL TERMS

Write a brief definition for each term.

1. Usage ______________________________

2. Production volume ______________________________

1. Usage refers to the extent to which the engineer's solution will be employed—such things as how many people will travel across it per year, how many times it will be run in its expected life, how many miles it must travel without major overhaul, and so on.

2. Production volume refers to the number of units of the engineer's design that are to be manufactured or built—such things as how many machines will come off an assembly line run, how many plants like this will be built, and so on.

Problem Definition—Some Recommendations

Don't get caught solving a problem you have failed to define! Furthermore, define a problem in two stages, beginning with a broad, detail-free *problem formulation* and following it with a *problem analysis* in which your general statement of what is wanted is translated into specific problem characteristics. This two-stage definition counteracts the natural tendency to become immediately enmeshed in details, after which a broad perspective is virtually impossible to achieve.

Problem formulation is a crucial stage in an engineering project. It is here that the thrust of the project is determined. At this point you are deciding what problem you are going to solve. What more important decision can there be in the entire problem-solving process? What belies its importance is the fact that it may take only a matter of minutes.

Problem analysis, in contrast, is relatively time consuming, since it ordinarily involves a great deal of consultation, observation, thought, fact gathering, and negotiation over restrictions. Incidentally, the treatment of problem analysis in this unit is strictly introductory. Furthermore, the definition of terms is loose.

YOU DESIGN IT

As you study the following interoffice memo from Designwell Associates Chief Engineer, James C. Hogan, keep in mind that you will be asked to respond with

1. Several alternative formulations of the problem, and

2. An analysis of the problem patterned on the model in Figure 6.

DESIGNWELL ASSOCIATES

Consulting Engineers Inter Office Memo

TO: Stu Dent, Project Engineer

FROM: James C. Hogan, Chief Engineer

The following request for a design proposal was received from Juicy Fruit Orchards, one of the largest growers of apples. I quote:

"We are seeking significant improvements in our apple harvesting methods. A severe labor shortage and rising labor costs have prompted us to seek proposals for harvesting systems that require less labor. To be competitive with current hand-picking methods, the harvesting cost of an alternative system cannot be greater than $.20 per bushel."

"Juicy Fruits harvests approximately 900,000 bushels of apples that are converted to a variety of apple products at a nearby processing plant."

"Most of our apple trees are between 4.5 and 6.0 meters tall, and an average of 6.3 meters in diameter. Apples are removed from the orchard area in large pallet boxes, 112 x 112 x 81 (height) centimeters. Not more than five percent of the apples sold whole for eating purposes may be bruised. Not more than thirty percent of the processed apples may be bruised. When harvested, the tree is stripped in one picking. The bulk of the harvesting is done between mid-September and mid-November."

You are assigned to this project. I expect a preliminary report in the very near future.

Figure 7

Formulate the problem assigned in this memo. In fact, give several alternative formulations. Consider the relative merits of each, and decide upon one to be used in your analysis form. Your formulations may be in words, diagrams, or sketches.

INPUT (STATE A) ——————→ OUTPUT (STATE B)

INPUT (STATE A)	OUTPUT (STATE B)

DEFINITION OF THE APPLE HARVESTING PROBLEM

FORMULATION

State A (Input) in General Terms ____________________

State B (Output) in General Terms ____________________

ANALYSIS

Specifications of states A and B in adequate details

State A ____________________

State B ____________________

Restrictions ____________________

Criteria ____________________

Volume ____________________

Usage ____________________

Solution variables ____________________

ALTERNATIVE FORMULATIONS

State A	State B
Trees laden with apples	Trees stripped, apples in boxes
Trees laden with apples	Trees stripped, apples at the processing plant
Trees laden with apples	Trees stripped, apples processed

These are three of many possibilities. The second makes most sense to us. The first is unnecessarily restrictive—the apples need not be placed in boxes. The third extends beyond the designer's authority.

FORMULATION

DEFINITION OF APPLE HARVESTING PROBLEM

State A in General Terms	State B in General Terms
apple-laden trees	trees stripped, apples at processing plant

ANALYSIS

Specifications of states A and B in adequate details

State A:

1. Apple-laden trees

State B:

1. Same trees, apple-free
2. Apples at the processing plant in suitable-to-handle containers

Restrictions:

1. Picking must be done in two month period
2. Tree must be stripped on one pass
3. Must use less labor
4. Cost per bushel $\leqslant$\$.20
5. Must bot bruise more than thirty percent of the apples

Criteria

1. Cost of installing, operating, and maintaining the system
2. Reliability
3. Safety
4. Labor required, both in hours and physical demands
5. Percent of fruit damaged

6. Adaptability to variation in tree size and shape, apple size and shape, and terrain
7. Vulnerability to inclement weather

Volume: One system

Usage: 900,000 bushels per year

Solution variables:

1. Method of separating apple from tree
2. Method of getting separator to apples
3. Method of collecting separated apples
4. Method of transporting to processing plant
5. Power source

SELF QUIZ

1. Define the term *state A*. ______________________________

____

2. Define the term *state B*. ______________________________

____

3. Define the term *criterion*. ______________________________

____

4. Define the term *restriction*. ______________________________

____

5. The term *input* is the equivalent of what other term? ______________________________

____

6. The term *output* is the equivalent of what other term? ______________________________

____

7. Define the term *black box*. ______________________________

____

8. Define the term *constraint*. ______________________________

____

9. Why is *usage* something that an engineer must be concerned with?

____

____

10. Why must an engineer know the production volume of a solution?

____

____

____

1. State A is the state of affairs in which a problem arises for a designer.
2. State B is a state of affairs that the designer seeks to attain.
3. A criterion is a standard by which a solution will be measured and evaluated.
4. A restriction is a condition that a solution must meet to be at all acceptable.
5. The term input is the equivalent of the term state A.
6. The term output is the equivalent of the term state B.

7. The term black box refers to a diagrammatic element in the formulation of a problem. The process of solving the problem is represented with no details—simply as an unknown step between input and output.

8. Constraints are the parameters within which variations in input or output must remain.

9. Usage is of concern to the engineer because the more a solution will be utilized—the more people who will use it, the more money saving operations that it will perform, and so on—the more its cost can be amortized over many units of utilization. In other words, the less it will cost per unit of useful employment.

10. Production volume greatly determines the practicality of expensive, hand-fabricated component parts, as opposed to inexpensive (per unit over many units), manufacturable components. It would be too expensive to set up a manufacturing line for turning out components if only a few machines were to be made. But it would be even more expensive to make components by hand processes if very many machines were to be manufactured.

Learning Segment 4
SEARCHING FOR SOLUTIONS

LEARNING OBJECTIVES

- Recognize that everyone's inventiveness is affected by many factors.
- Use solution-finding methods that will help overcome false limitations imposed on possible solutions.
- Use both random methods and more systematic methods of searching for solutions.
- Perform the search stage of the design process.

INTRODUCTION

A vast store of technical information does not, by itself, make an engineer valuable to employers and society. The payoff comes when an engineer applies this knowledge—solves problems—or more specifically, creates structures, machines, devices, or processes that satisfy human needs and wants. So, since engineering is basically a creative activity, a person who is prepared for it has read and thought about creativity and invention and has worked at developing his inventive skills. You can begin such efforts right now by carefully considering the contents of this learning segment. It is intended to focus your attention on the nature and importance of the creative act in engineering, to inspire you to enhance and capitalize on your creative abilities, and to show you how to do both.

Solutions occur to you quite unpredictably when you are laboring over a problem. You may get an idea the moment you learn about the problem, or during Sunday's sermon, or on your way to work, or three days after you've already "solved" it. This is the way of the mind; why fight it? But subconscious, leave-it-to-chance methods can hardly be relied on to solve consequential problems satisfactorily and consistently. What is called for is a deliberate and energetic search for alternative solutions in which conception of solutions is the objective rather than a by-product of your efforts. Therefore, after you have defined a problem, you must turn your efforts to what is truly a search of the mind, of the literature, and of the world about you—a search for solutions.

Surely you have wondered where a practicing engineer's solutions come from. Probably the first logical source to occur to you is books. True, man's vast accumulation of knowledge provides ready-made solutions for some parts of most problems. Searching for these is a relatively straightforward process of exploring textbooks, handbooks, technical reports, journals, existing practices and, of course, one's own memory. But there is a second major source of solutions—the designer's own ideas—the fruits of the mental process called *invention*. Any engineer *will* rely heavily on ingenuity to solve many aspects of problems not covered by existing technical and scientific know-how. Unfortunately, however, inventing solutions is not as straightforward and controllable as looking up ready-made solutions; you may recognize this from your own problem-solving experience. So it pays to devote special attention to improving your inventive ability.

For some people, learning that engineering work offers ample opportunity to exercise their ingenuity is a happy thought. It may make others apprehensive, but there is no need for this. Almost all people have—or can develop—what it takes to meet the creative demands of their jobs. That, of course, is an opinion, but a well-founded one. Hear our argument.

Your inventiveness depends on: inherited qualities; your attitude; your knowledge; the effort you exert; and the method you employ in seeking out ideas. This leads to a significant point. *Since you control all but the first of these five determinants, it is within your power to improve your inventive ability*. You *can*, over a period of time, develop a more favorable attitude and increase your knowledge. You *can* increase your effort. You *can* substantially improve your method of searching for solutions. This should be good news to you if you were not born a creative genius; it means you can compensate for this through your influence over the remaining four determinants. Some elaboration is in order.

CHECK IT OUT

Most people are a lot more creative than they think. Many times it is difficult to get started creatively when one is all alone. In such a case, a bull session of five to ten people can reveal a wealth of hidden creativity.

Choose a problem that a number of your fellow workers would be interested in and, during fifteen minutes over lunch, try to get five or more of them into a discussion of possible solutions. If you prefer not to use a problem from work, you may wish to try finding either uses of flywheels or perhaps ways of removing an oil slick from an open body of water.

Try to keep the discussion flowing with more and more possible solutions, no matter how crazy they might sound at first. Don't let negative reactions to wild ideas slow things down—keep everyone positive and active or the discussion will be a failure as a demonstration of hidden creativity.

You, of course, are the sole judge of the success of this exercise.

THE INFLUENCE OF ATTITUDE

You won't get far in your efforts to become more creative if you don't believe in yourself.

To succeed, you have to be convinced that you can be as inventive as an engineer need be. Hopefully, you will be convinced by the following argument.

Why are some persons noticeably more inventive than others? "Born that way" is what you are probably thinking. Hogwash! On what basis do you conclude that heredity accounts for this when it is only one of the five determinants of a person's creative performance? An unusually creative person could just as well be that way not because he was exceptionally endowed at birth but because his attitude, knowledge, or method is exceptional, because he works hard, or for any combination of these reasons. It may well be that a rare few are exceptionally endowed through heredity, but it is also true that everyone inherits inventive ability and that very few people fully exploit the potential they have. It is very likely, as far as this trait is concerned, that you weren't shortchanged at birth. So don't be concerned about whether you have aptitude for invention; concern yourself with putting what you have to use.

Develop a Hunger for Solutions

It will also help if you can develop an insatiable drive to find better solutions. Rarely, if ever, will you be involved in an engineering project in which you are able to identify all solutions. Usually you will run out of time long before you run out of possibilities. The engineer who developed a baggage handling system for an air terminal conceived various methods of transporting passengers' luggage, numerous sorting systems, and many pick-up arrangements. He finally ended his search because other problems were awaiting his attention, not because ha had exhausted all possibilities. Nor did he believe at any time during this project that he had thought of all solutions. He assumed that more and better ones remained, and he went after them. He is what we call "alternatives hungry."

Become Alternatives Conscious

If you are to be that way you must become alternatives conscious. One way of achieving this is to bombard yourself with problems to which there are impressive numbers of alternative solutions. (See Figure 8.) If this produces the desired alternatives consciousness, there is little chance you will be caught "solving" a problem without uncovering and considering a reasonable number of alternatives. Each time you isolate an additional solution, set out again to find a better one, continuing until you are forced to give up because of a project deadline or the pressures of other problems awaiting your attention.

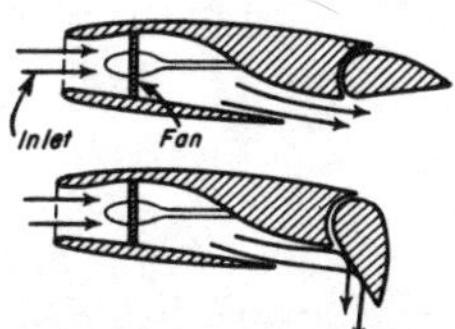

Figure 8. Speaking of alternative solutions, here are some of the methods devised for powering a plane in vertical flight. These, plus the familiar helicopter, are being considered for military and commercial use.

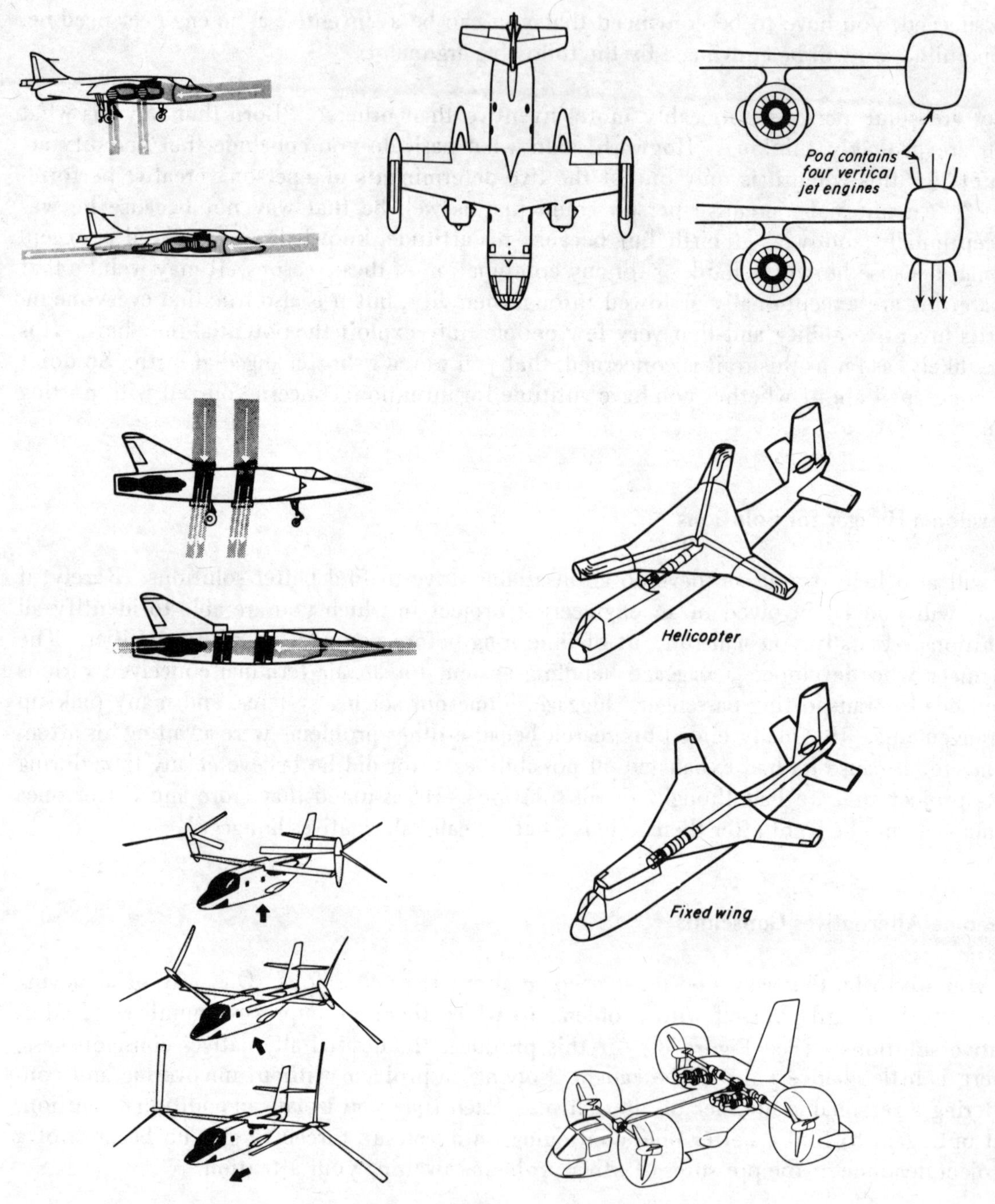

THE ROLE OF KNOWLEDGE

What we loosely refer to as an idea is the combining of two or more bits of information in a manner that is novel to the conceiver. It is a reorganization of knowledge; no idea evolves from nothingness. And so the larger your store of knowledge is, the more raw material you have from which you can synthesize solutions. Furthermore, the broader the range of subjects this knowledge embraces, the better the prospects for unique and often particularly effective ideas. School is not the only source of such knowledge; observation, conversation, reading, and other forms of lifelong learning are important, too.

CHECK IT OUT

It is sometimes helpful to take inventory of one's strengths and weaknesses in knowledge. We are then in a better position to do something about areas of weakness by making the effort to learn in those areas. Indicate your own strengths and weaknesses below.

Strong areas of knowledge ______________________________

____

____

Areas of knowledge that I would like to strengthen ______________________________

____

____

You have the answer on how to strengthen the latter areas—study.

THE INFLUENCE OF EFFORT

There is a big difference between an occasional flash of genius and consistently producing, under pressure, ideas and more ideas, to solve a given problem in the limited time available. The latter takes effort. You won't find many really creative persons who are not hard workers. So, to maximize your inventiveness, you must be willing to think long and hard, even to sweat a little.

CHECK IT OUT

I am basically satisfied with my level of effort in: ______________________________

____

____

I am not satisfied with my level of effort in: ______________________________

____

____

Two specific steps I can take to increase my level of effort in areas that I wish to improve are: ____

____

____

THE INFLUENCE OF METHOD

The following analogy highlights some of the pitfalls, difficulties, and procedural flaws you are vulnerable to when seeking solutions to a problem. Visualize the *X*s pictured in Figure 9 as points in space, each representing a solution to a problem at hand. Assume that widely

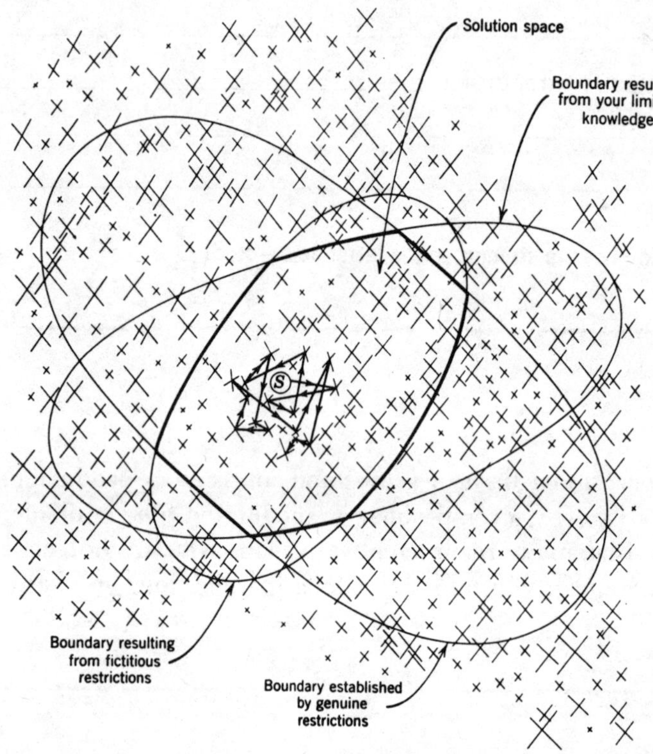

Figure 9. The arrowed path represents the manner in which you are prone to search for new solutions to a problem unless you have devoted careful attention to your search method.

separated points represent radically different solutions. When trying to think of solutions to a problem, wouldn't it be nice if you could start somewhere in this solution space and move to progressively better solutions until perfection or, more likely, a deadline ends your search.

You may be thinking, "Not I." Yet you would probably be appalled if you carefully and objectively examined what you actually do when seeking ideas. All too often you start with the presently used solution and proceed from one point (idea) to another in a manner indicated by the arrowed path. Notice that the jumps tend to be relatively small, so that ideas tend to cluster about the current solution.

Why do solutions tend to closely resemble the present way(s) of solving the problem? Mainly because the customary solution to a problem has an almost uncanny power of attraction. This power is especially strong if that solution has a long history of use. After you have "lived" with a certain solution for a while and have become so accustomed to it, your thinking is bound to be in a rut.

THINK IT THROUGH

Is it really true that attempts to solve problems tend to cluster around previous solutions? Can you think of any examples from history?

____

____

____

One answer to this exercise appears in the following paragraph.

History bounds with illustrations, an excellent one being the record of man's attempts to fly. The solution men were familiar with was the wing action of birds and insects. Almost inevitably, then, they tried to fly be employing all sorts of unsuccessful and often disastrous wing-flapping schemes. In time they freed their thinking from "the stranglehold of the familiar."

This analogy will help you to appreciate why it pays to devote careful attention to your method of searching for solutions. It indicates a few of the tendencies that inhibit your ability to invent, unless you take measures to minimize their effects. Fortunately, such measures exist. Give them serious consideration; they are probably your best bet if you wish to improve your inventiveness. They are:

- *First, maximize the number and variety of solutions on which you can draw by enlarging the solution space as far as practical.* You do this by pushing back the boundaries that establish that space. There are opportunities to do so in almost every problem: by ridding yourself of imagined and self-imposed restrictions, by shaking off unjustified "real" restrictions, and by supplementing your knowledge relevant to the problem at hand.

- *Then, take full advantage of this expanded solution space; search it effectively.* You should sample all areas of possibility that offer promise of containing the optimum solution.

This requires that you triumph over any number of temptations. We can help you with a strategy found to be very effective: making your search partially systematic and partially random. Both approaches warrant elaboration.

Adding a Semblance of System to Your Slarch

An excellent way of doing so is illustrated here. An engineer is engaged in the design of an improved system for harvesting apples. In the course of defining the problem, he identifies solution variables such as "method of separating apples from the tree," "method of bringing the separator (for example, a man) to the apples," and "method of collecting the separated apples." As he concentrates on the solution variable "method of separation," he searches first for basic methods of separation; then he looks for specific versions of each of these basic alternatives, as illustrated by Figure 10. This minimizes the chance that he will overlook a whole block of promising possibilities. He then concentrates on other solution variables in turn, attempting to accumulate as many possibilities for each, always working from the general to the specific. It is doubtful that you will find a more effective method of improving your inventiveness.

The alternative solutions generated for a solution variable (for example, those in Figure 12 for "method of separation") are referred to as *partial solutions*. Eventually the engineer will evaluate these partial solutions and combinations of them, perhaps recombining and re-evaluating numerous times, until he has synthesized a complete solution that is the best combination of partial solutions.

YOU DESIGN IT

Use the space that follows, and a few minutes that you can steal from somewhere, to work out some partial solutions for your apple harvesting design problem. You will want to refer to your notes at the end of the last learning segment, but please do not look ahead to Figure 12 before you have tested your own ingenuity.

There are other ways of introducing system into your search. You can systematically concentrate on each criterion, trying to generate, for example, means of maximizing reliability or of minimizing construction cost. You can systematically rearrange partial solutions in hopes of finding a high-payoff combination. The *alternatives tree* (Figure 12) is an effective means of systematizing your thinking. Any other way of organizing your thoughts and investigations so that a wide range of basically different solutions is brought under consideration is likely to be profitable.

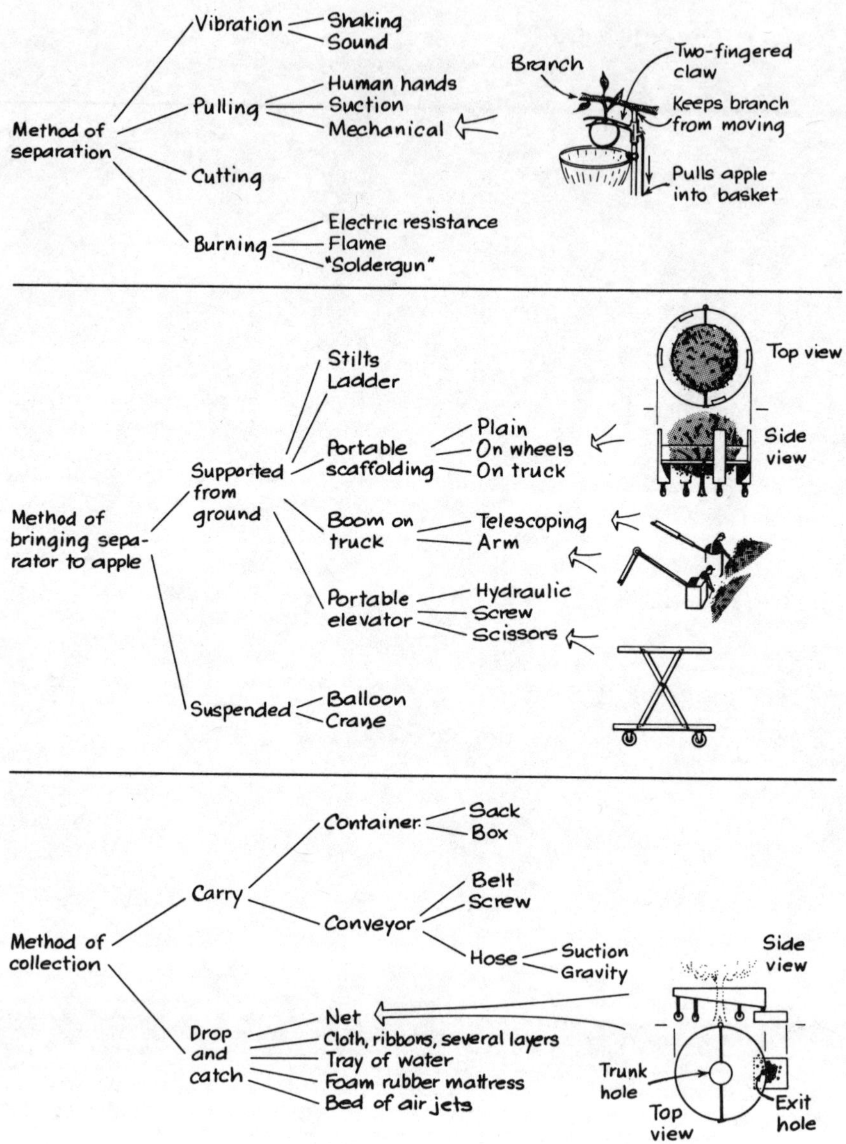

Figure 10. A page from the engineer's notebook, showing some of the alternative partial solutions he generated for the apple-harvesting problem. For each major solution variable, he classified his ideas in the form of an alternatives tree.

TECHNICAL TERMS

Write a brief definition for each term.

1. Partial solution ______________________________

____

---

2. Alternatives tree ____

---

1. A partial solution is a possible alternative method or set of specifications to satisfactorily overcome the problem represented by one solution variable. For instance, specific shapes for the solution variable of shape, or a specific dimension for the solution variable of width.

2. An alternative tree is a diagrammatic means of summarizing the partial solutions for a solution variable. In this case an example does so much more than words, so review the three alternatives trees in Figure 10.

Random Methods

There are some predominantly random methods of getting the mind into what might otherwise be "unexplored territory." Notable is the technique of *brain-storming*. A half-dozen or so persons assemble for the purpose of generating solutions to a problem. The leader describes the problem and the participants then bombard him with ideas, which he records on a blackboard. The objective is to accumulate many ideas by creating an atmosphere that encourages everyone to contribute all solutions that come to mind, regardless of how absurd they may seem at the moment. Ridicule and other negative responses are penalized. No attempt at evaluation is made during the session; this comes later, and wisely so. After some experience with this technique, a group can achieve a freewheeling flow of ideas that yields an impressive number and variety of solutions, often in 15 minutes or less.

One reason that brainstorming is fruitful is that the rapid-fire flow of ideas is repeatedly redirecting each participant's thoughts into new channels. In terms of the model pictured earlier (Figure 9), each person's mind is buffeted about the solution space in random fashion, forcing "big jumps" to distant points, thus combating the clustering tendency. The probability is high that some of these excursions will lead to profitable ideas.

Now you know what is meant by a "partially systematic, partially random search strategy." It is intended, one way or the other, to guide your search for solutions into profitable areas of possibility you might otherwise overlook. Since systematic methods have their limits and since you certainly do not want to rely only on a random search, it pays to call on both approaches.

TECHNICAL TERMS

Define what is meant by a "partially systematic and partially random search strategy."

A "partially systematic and partially random search" relies both on chance to uncover solutions (as seems the case when ideas strike us at odd moments) and on organized efforts, such as the alternatives tree, checklist, and literature search.

YOU DESIGN IT

Identify one major solution variable for the problem of snow removal from city streets, and generate as many partial solutions as you can for the variable.

VARIABLE PARTIAL SOLUTIONS

Some major solution variables for snow removal, with some alternative partial solutions for each:

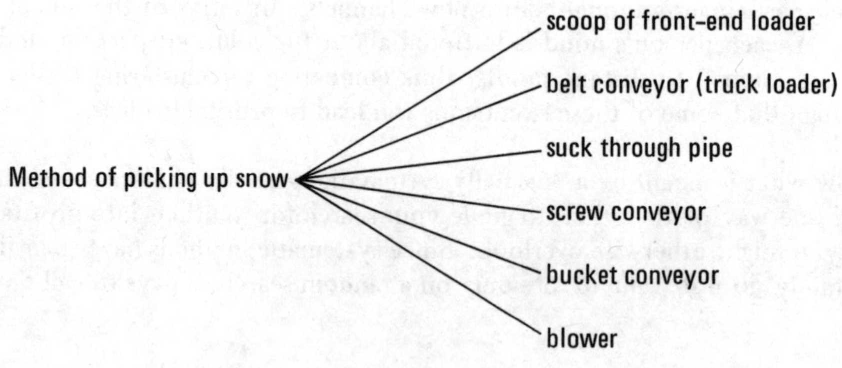

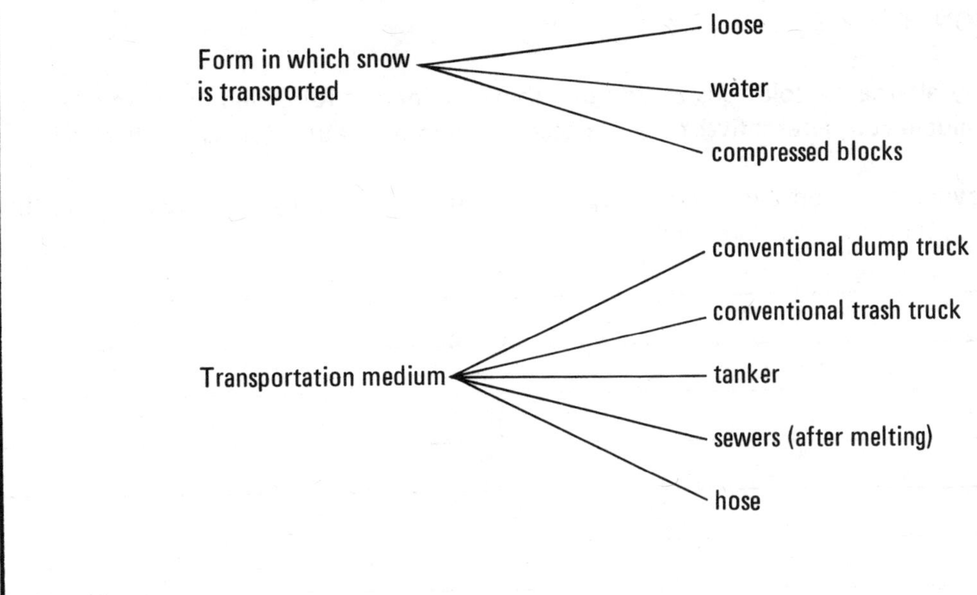

TWO IMPORTANT DON'TS

Don't get bogged down with details sooner than necessary. Suppose you start working out the details of the first "good" idea you have. For all practical purposes your search will probably end right there. You will be spending time on details when you should be searching for other basically different solutions. Furthermore, preoccupation with the details of one solution severely hampers your ability to think of significantly different ones. Also, if you fall prey to this temptation and you happen to uncover a superior solution later, you will probably be unjustly biased in favor of the solution in which you have already invested so much time on details. And, finally, many alternatives can be satisfactorily evaluated while still in a relatively crude state of specification; since most will be rejected, why waste time on detailing them?

Therefore, postpone details until they become necessary for decision-making purposes. In fact, it is best to form only solution concepts in this phase of the design process. (A *solution concept* is the essence, the gist, the general nature of a particular solution. Its form may be a rough sketch, a few words, or a sentence or two. The ideas shown in Figure 10 are partial solution concepts.)

Avoid premature evaluation; it has the same detrimental effects as premature preoccupation with details. This is the *search* phase of the design process; it is followed by the *decision* phase, in which evaluation of alternatives predominates. Therefore, ideas will not go unevaluated, but good ideas may go undiscovered if you become preoccupied with evaluation when you should be searching for better solutions.

YOU DESIGN IT

List as many alternative solutions as you can think of for the following problem. Do not stop and evaluate your alternatives or get bogged down in premature concern with details.

Cartons moving on a conveyor belt are spaced irregularly. As they approach the labeling machine, they must be spaced at regular intervals.

To get a set of answers to compare with yours, you will have to do some library research. See the March, 1965, issue of Modern Materials, which has an excellent spread on alternative solutions.

Enough talk about techniques for improving your inventiveness. Perhaps you are convinced by now that there are things you can do about it. Certainly much more could be said about the process of invention, particularly its stimulation, and much more *has* been said. It is a popular topic for books, articles, and papers, so dearth of literature is certainly no obstacle if you wish to explore this subject further.

A PLEA FOR MORE ORIGINALITY

These parting comments on the search phase concern what has become a cause for us: promotion of more imaginative thinking on the part of engineers. Too many solutions are the offspring of the past; too few are the products of original thought. Inertia perpetuates a host of inferior solutions in the world about us, leaving lots of running room for more inventive engineers.

THINK IT THROUGH

Can you think of any possible reasons for engineers to rely unduly on solutions of the past?

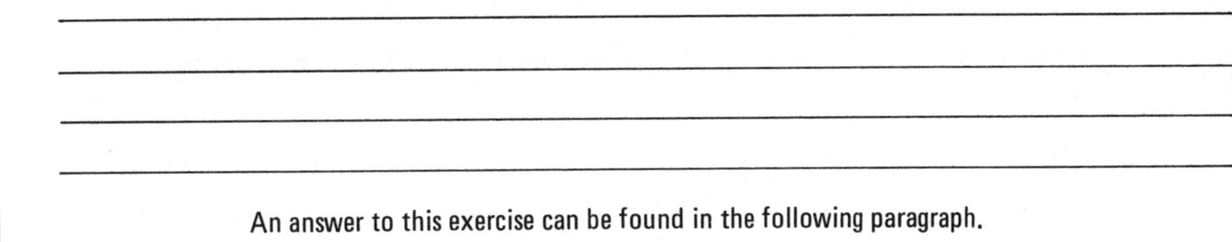

An answer to this exercise can be found in the following paragraph.

There are explanations for this undue reliance on solutions from the past. It is certainly the path of least effort. Moreover, habitual solutions, although imperfect, are at least proven, whereas the performance of radical departures can be discomfortingly uncertain. Furthermore, you are likely to find that your problem-solving efforts become less imaginative as you become more knowledgeable in your field of specialization. It's a fact of life—the more you learn through education and experience about the accepted manner of doing things, the more constrained your thinking becomes. This is why some of the most imaginative thinking in a field often originates from people who are new, sometimes virtual foreigners to that discipline. How often we have assigned a simple engineering problem to freshmen engineers only to have them moan, "How can we solve a problem like this? We don't know anything about it." Little do they realize that their ignorance is in their favor as far as originality is concerned. They usually produce some imaginative solutions; they *can't* fall back on the customary.

It is easy to miss the point here. No one is knocking the vast store of valuable know-how that man has accumulated. The point is this: by all means acquire specialized knowledge, use handbooks, learn through experience how things are done. But draw on this store of customary solutions *and* on your capacity for original ones; both have much to contribute. The trick is to be creative in spite for your specialized knowledge, since without that you can't get along in modern engineering.

You can derive more than the obvious from the "creativity message" about which you have been reading. In particular, you should have gained further insight into the practice of engineering, the need for creativeness in engineering, and the attention given to it by engineers.

For years I have been promoting the idea of a campus seminar on the subject of creativity. In it architects, artists, musicians, poets, authors, scientists, mathematicians, engineers, and others would exchange views of creativity, its nature, enhancement, and perception. All engineers are artists to an extent, and surely an exchange on this subject would be a stimulating experience. But enough of this vision; the point of mentioning it is the reaction I get from my nontechnical colleagues. They are obviously surprised, and can't quite understand why an engineer is interested in creativity. Nor can they envision what engineers might contribute to such a seminar. Such responses reveal one of the common false impressions of engineering.

SELF QUIZ

1. List the five things that were mentioned as having an influence on your creativity or inventiveness.

2. Explain what a boundary is in relation to a solution. Give some examples of kinds of boundaries.

3. Explain what is meant by *expanding your solution space*.

4. Define a *partial solution*.

5. Define an *alternative tree*.

1. Inherited qualities, attitude, knowledge, effort, and method

2. A boundary is some limitation that is imposed on the number, kind, location, or other aspect of a possible solution to a problem. It prevents the designer from searching in any areas blocked by these limits or considering any possible solutions introduced from outside them. Boundaries include such things as "that's the way we've always done it," ignorance or lack of knowledge, fascination with previous solutions, real restrictions im-

posed by the nature of the problem or by the parties that will utilize the solution, fictitious restrictions that are imposed by false statements of the problem, and so on.

3. Expanding one's solution space involves ridding oneself of limits as to what solutions may be considered. It requires overcoming false and self-imposed restrictions, fighting off unjustified restrictions imposed by others, increasing one's knowledge, and so on.

4. A partial solution is a possible alternative method or specification that overcomes the problem represented by a solution variable.

5. An alternatives tree is a diagrammatic means of summarizing the partial solutions for a solution variable.

Learning Segment 5
EVALUATING ALTERNATIVES AND DECIDING ON A SOLUTION

LEARNING OBJECTIVES

- Understand the meaning of such terms as *benefit-cost ratio, unquantifiables, human engineering, usability, safety, simplicity,* and *elegant solution*.

- Know some of the ways that engineers and designers anticipate pitfalls that would lessen the safety or usability of their creations.

- Understand and perform the four steps in the decision process—

    - Select and weight criteria.

    - Predict the performances of alternative solutions.

    - Compare predicted performances.

    - Make a design decision.

THE GENERAL DECISION-MAKING PROCESS

In the search phase an engineer expands the number and variety of solution candidates. What is needed next is an elimination procedure that will reduce these alternatives to the preferred solution.

For example: An automobile manufacturer had for some time been using an expensive machine to mount tires on wheels. Because of an increase in the number of mountings to be made, an engineer was asked to do something to speed up the process. The result of his efforts was a new and much simpler device (Figure 11). Before proposing this device to the management, he made a thorough comparison of his solution and the existing one. The results are shown in Figure 12, which shows the criteria on which he based his decision, how the alternatives stack up with respect to these criteria, and the tabular summary he prepared to facilitate comparison. This case is unusual in that it involves only two major alternatives, but its uncomplicated nature makes it an excellent illustration for the following introduction to decision making in engineering.

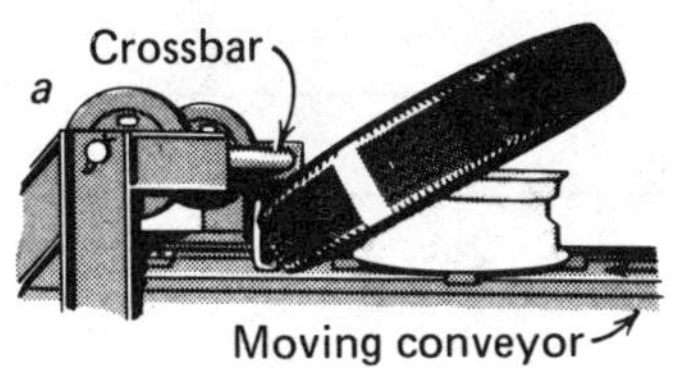

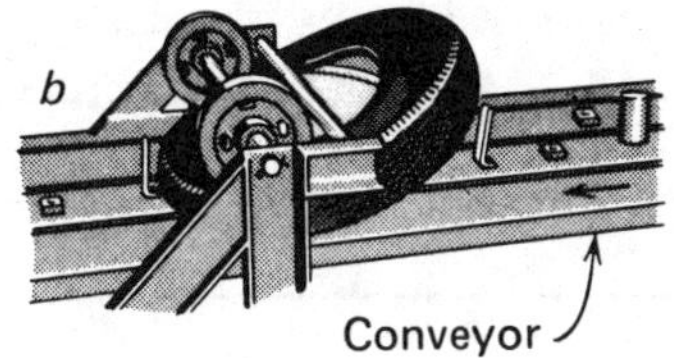

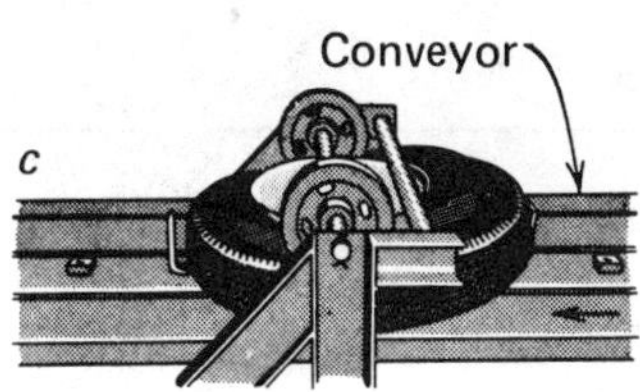

Figure 11. (a) The beads of the tire are lubricated to insure that they will slip over the rim. The tire is on the wheel at an angle of approximately 30° to the rim as it approaches the cross bar. (b) As the tire moves under the cross bar it is engaged by two wheels located parallel to the centerline of the conveyor. These wheels are the only moving parts of the new machine; they squeeze the leading beads of the tire into the drop center of the wheel. (c) The remainder of the tire is forced onto the wheel by the cross bar.

Criteria		Proposed Device	Present Machine
Investment	Cost to build, install, "debug"	$750	0
Operating expenses (for 5 years)	Attendant	$600	$23,000
	Maintenance	350	900
	Repair	250	750
	Power	0	400
	Total	$1200	$25,050
The unquantifiables	Reliability	Better	
	Safety	Better	

$$\frac{\text{Benefit}}{\text{Cost}} = \frac{\text{Saving in operating expenses}}{\text{Cost of building \& installing}} = \frac{\$25{,}050-\$1200}{\$750} = 31.8$$

Figure 12. This is a page from the engineer's final report, comparing the performance of his device and that of the one presently used. He predicts that $750 will be required to build, install, and "debug" his device. The costs of an attendant, maintenance, and the like have also been predicted for the five years that he expects his device will be needed. He acquired equivalent data for the existing machine and summarized these costs in this table. He also calculated a benefit-cost ratio that certainly helps his superiors appreciate the merits of this $750 investment and you to understand why they adopted it so readily.

Although the specifics vary from situation to situation, in almost every instance the following steps must be taken before an intelligent engineering decision can be reached:

1. Criteria must be selected and their relative weights determined.
2. Performance of candidate solutions must be predicted with respect to these criteria.
3. The candidates must be compared on the basis of these predicted performances.
4. A choice must be made.

THINK IT THROUGH

Which criterion do you think carried the most weight in the decision to adopt the solution shown in Figures 11 and 12? Why?

An answer to this exercise appears in the following paragraph.

Selecting and Weighting Criteria

Usually the overriding criterion is the *benefit-to-cost* ratio, which is the benefit expected from a solution relative to the cost of creating and using it. For the tire mounter the *benefit* is the saving in operating cost made possible by the proposed device. The *cost* is the total expense incurred in building and installing it (bottom of Figure 12). The benefit-cost ratio being employed in deciding on a proposed dam is

$$\frac{\text{Flood protection} + \text{conservation and recreation benefits} + \ldots}{\text{Cost of land} + \text{construction cost} + \text{maintenance cost} + \text{hardship to displaced persons} + \ldots}$$

The people who are footing the bill are interested in the expected tab as well as the prospective benefits of a proposed engineering venture. In fact, claims of benefits mean little if the associated costs are unknown. Knowing that a certain machine will cut his harvesting costs by X dollars a year does not mean much to a farmer until he also knows the cost of that machine. As a taxpayer, you insist on being told the cost of a proposed public works project before passing judgment on it. Similarly, an engineer rarely describes the benefits attributable to a proposal without also quoting the cost of bringing about those benefits.

A term used synonymously with benefit-cost ratio is *effectiveness-cost ratio*. Furthermore, some authors speak of cost-benefit instead of benefit-cost, and others talk of *return on the investment* but, regardless of the specific terms or their ordering, the notion is the same important one: total benefits relative to and inseparable from total cost.

TECHNICAL TERMS

Write a brief definition for the term *benefit-cost ratio*. Also, give at least three synonymous terms.

A benefit-cost ratio is the total expected benefit to be derived from implementing a given solution. The benefit is figured in relation to the total costs expected to be incurred in creating and implementing the solution. Synonyms include cost-benefit ratio, effectiveness-cost ratio, and return on investment (ROI).

The Recurring Question of Economic Feasibility

There are remarkably few things that man cannot achieve, given sufficient time and money. Whether a device can be developed for a given purpose is seldom the question. It can almost certainly be done if someone is willing to pay the price. The real question usually is, can a solution be developed profitably? Therefore, in most instances, technical feasibility is not an issue; economic feasibility definitely is.

Implicit in the inception of a design project is the assumption that a solution to the problem is economically feasible, that investment of engineering and other resources to develop a solution will be repaid with a profit. Before starting to design a machine for assembling reed switches, the engineers deliberated over the prospects of a profitable result. The decision to proceed was based on their judgment, on past experience in similar ventures, and on a willingness to accept a certain risk.

From the moment a project is begun, the hypothesis that a profitable solution will evolve is under test. This question is being asked repeatedly, although not always explicitly, through the design process: "On the basis of what has been learned thus far in the project, are the prospects of the venture yielding a satisfactory return on what apparently must be invested sufficiently high to justify continuing?" Thus, at any time from its inception to the specification stage, a project is subject to termination if the information accumulated indicates that a profitable solution probably will not be found under the current state of technology.

Of course, at the outset of a design project, relatively little is known about the probable final solution, so that at this stage there are many uncertainties and thus a sizable risk of being mistaken in assuming that the venture is economically feasible. As the project progresses, alternative solutions evolve, possibilities are appraised, and other information on which to base an answer to the ever-present profitability question becomes more substantial. Therefore, the risk of making the wrong decision is maximum at the start of a project and decreases thereafter as progress is made and evidence accumulates.

Making recommendations concerning economic feasibility of proposed projects is a very important part of an engineer's work. Such decisions are far from simple, and yet the engineer who acquires a record of blunders in this respect is not favorably regarded.

THINK IT THROUGH

Normally, one cannot simply arrive full blown at a benefit-cost ratio. What preliminary steps are required to arrive at an overall benefit-cost ratio?

An answer to this exercise appears in the following paragraph.

Ordinarily, to estimate the benefit-cost ratio satisfactorily, a number of subcriteria must be evaluated first. Subcriteria that make up total benefit and total cost of the proposed dam are apparent in the preceding ratio. In aggregate, these subcriteria determine the benefit-cost ratio.

Predicting Performances of Alternative Solutions

Predicting how well each alternative will fare if it is adopted is the most demanding part of the decision-making process. For the proposed tire-mounter, the engineer had to predict the cost of building it, the number of man-hours needed to maintain it, its reliability, and so on. To make these predictions, the engineer relied mainly on his judgment and on experiments with a working version of his proposal. (The engineer has other means of predicting performance, such as mathematics and simulation, which are described elsewhere.

Predicted performances should be in the same units if they are to be accumulated and compared intelligently.

THINK IT THROUGH

It is difficult if not impossible to compare the predicted performances of alternative solutions if they are not measured in the same units. What do you suppose is the most convenient unit to translate performance into?

An answer to this exercise appears in the following paragraph.

By far the most convenient measure for these purposes is money. Thus, Figure 14 shows predicted performances in dollars for the criteria for which this is feasible. There will always be some criteria that defy numbers. Note that although it was too expensive and time-consuming to measure safety and reliability in this case, the engineer did not ignore them. He judged them in qualitative terms and concluded that they reinforced the monetary argument for this device.

Comparing Predicted Performances

To make an intelligent choice from the alternatives, they must be compared meaningfully with respect to the criteria. When dealing with criteria for which monetary predictions are feasible, these figures are usually tabulated or otherwise aggregated so that costs and benefits can be easily compared. Incidentally, this example, simplified for our purposes, illustrates one of a number of methods of making economic comparisons of engineering alternatives. These procedures derive from a rather extensive body of knowledge of fundamental importance in engineering, generally referred to as engineering economics.

YOU DESIGN IT

Can you picture it? You are an engineering consultant to a grower of thousands of acres of tomatoes. Your mission is to develop a machine that will significantly reduce harvesting cost. You now have preliminary designs. They are: (a) machine-assisted handpicking, and (b) completely mechanized harvesting which, of course, requires a larger investment but has a lower operating cost than alternative *a*. What criteria are you going to apply in choosing between these? How do you predict the alternatives will fare with respect to each of these criteria?

Criterion 1 ____

____

Criterion 2 ____

____

Criterion 3 ____

____

Outcome of alternative *a* on criterion 1 ____________________________

____

Outcome of alternative *b* on criterion 1 ____________________________

____

Outcome of alternative *a* on criterion 2 ______________________________

____

Outcome of alternative *b* on criterion 2 ______________________________

____

Outcome of alternative *a* on criterion 3 ______________________________

____

Outcome of alternative *b* on criterion 3 ______________________________

____

Criteria: Development and construction cost, energy required, labor cost, physical exertion involved, worker skills required, availability, maintenance cost, effects on the environment, damage to tomatoes, percentage of crop recovered, vulnerability to weather, hardship caused through unemployment, and safety.

Alternative a: machine-assisted picking is likely to: have a higher labor cost, require less in worker skills, be more demanding physically, be more vulnerable to inclement weather, be more dangerous, and be unavailable a greater percentage of the time (it is vulnerable to mechanical failures and strikes).

Alternative b: "complete" mechanization (no pickers are required but there is an operator) is likely to: cost more to develop and construct, require higher-skilled personnel for operation and maintenance, consume more oil, be more of an environmental nuisance, do more damage to tomatoes, leave more of them on the vines, and cause more unemployment.

Making the Decision

The thoroughness and rigor with which these steps are executed vary considerably from problem to problem, depending on the complexity of the solutions being evaluated, their competetiveness, the importance of the decision, and other circumstances. In one situation the most elaborate, exhaustive procedures, involving much measurement, modeling, and cost investigation, are warranted. In another, only a quick, simple, informal judgment is justified. The elaborateness of the evaluation process also depends on the stage of decision making. At the outset there are likely to be many competitors at least some of which, even though only vaguely conceived, can be eliminated by quick and simple evaluations. The remaining candidates, probably with details added, warrant more refined evaluations. As this multistage screening process continues, the competitiveness of the remaining candidates, the detail to which they are specified, and the elaborateness of the evaluations will increase.

YOU DECIDE

Many a driver has jockeyed at 95 kilometers per hour to get into the left lane only to discover a half-mile later that the message this sign was meant to convey was, "Be in right lane if you intend to turn right on Route 50." Surely you can tell highway engineers how to accomplish this, and surely you can do it simply. Indicate your approach in the space below.

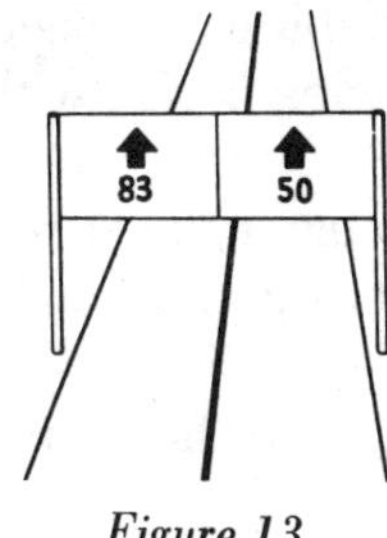

Figure 13

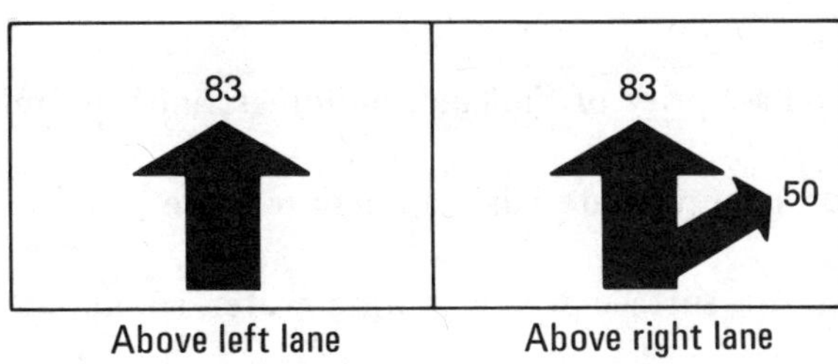

Figure 14

MORE ON CRITERIA—THE ELUSIVE ONES

Only a minority of criteria can be simply and accurately measured. Most likely, in a given project, more than half of the criteria to be considered are either impossible or impractical to measure, or in some other respect are troublesome for decision makers. Considering the difficulties these "troublemakers" cause, the risks associated with ignoring them, and the vulnerability of engineers in such matters, it is well worth discussing them at length. This discussion begins with a case in point.

A young engineer has been hired as a technical consultant to assist a city administration in the selection of a solid-waste management system. No doubt about it, there are alternatives–recycling, conversion to energy, and landfill, to name a few. In the course of his work for the city he naturally prepared a list of criteria on which the choice of system should be based.

THINK IT THROUGH

You should be able to anticipate at least a half-dozen of the criteria listed by this young engineer. Try.

____

____

____

____

____

____

A baker's dozen from the young engineer follow.

The criteria he proposed are:

1. *Initial investment*. Purchase price of the land, buildings, and equipment.

2. *Operating costs*. Labor, energy, materials, and maintenance.

3. *Environmental effects*. Air, surface-water, ground-water, sound, and visual impacts.

4. *Availability*. It's important; it depends on frequency and duration of shutdowns for repairs, maintenance, and cleaning.

5. *Safety*.

6. *Vulnerability to changing regulations*. Governments have the nasty habit of periodically tightening the ground rules under which landfills, incinerators, and other solutions must be operated, and this often ups the cost of operation.

7. *Public acceptance.*

8. *Adaptability to new materials.* Alternative methods differ in their ability to accomodate new materials that are sure to come along. Landfill accepts everything, but others do not.

9. *Predictability.* There *are* varying degrees of uncertainty associated with different methods (for example, the unpredictable markets and prices for recycled materials).

10. *Expandability.*

11. *Skill required for operation.* It's of consequence not only because of wages. Some methods require a greater percentage of highly-skilled people, which yields not only an administrative headache but a different ratio of skilled to unskilled jobs on the public payroll.

12. *Energy consumption.* This is over and above the direct cost of energy required; here it is a matter of consuming versus conserving a societal resource.

13. *Proportion of materials conserved.* Like criterion 12, this is beyond immediate economic concern.

THINK IT THROUGH

A careful examination of this young engineer's list of criteria should result in a number of observations about criteria coming to your mind. Jot down a few of the more important ones now. Then read the author's observations.

____

____

____

____

Some of our observations follow.

Here are some observations we hope you made as you examined his criteria.

- That the selection and weighting of criteria are consequential matters in an engineering project.

- That there are more criteria to be considered in such a decision than many people surmise. The number in this instance is not unusual and, to be sure, it complicates things.

• That some criteria, like initial investment, can be evaluated in dollar terms and with relative ease.

• That others, like safety and public acceptance, are impossible or impractical to predict in numerical terms and therefore can only be considered qualitatively in the decision process. Engineers like to refer to these as the *unquantifiables* or *irreducibles*. These, indeed, complicate decision making.

• That there is considerable uncertainty inherent in the prediction of future legislation, new materials, and public acceptance. The result is that a surprising amount of "crystal balling" is required, which is always a source of a certain uneasiness.

• That a majority of these criteria involve consequences directly and only for the municipality using this solid waste management system. But there and in many other engineering undertakings nonusers are also affected. Remember, other communities, downwind and downstream, that do not benefit from this system still suffer the pollution consequences.

• That some of these criteria represent a long-term concern for the city's welfare. A criterion like expandability implies caring about possible expenditures perhaps 25 years hence.

• That some criteria (for example, energy consumed and material resources destroyed) require a concern that extends beyond city limits and beyond the immediate future. They imply caring about the material and energy resources available to future generations.

• That all this adds up to a challenging decision-making situation because of the number of criteria, the conflicts between them, the predominance of unquantifiables, and the uncertainties involved.

This engineer did his job well. His list of criteria reflects no neglect of unquantifiable criteria, no disregard for indirect effects, no blindness to the consequences for nonusers, and no indifference to long-term implications. And that's the way it should be.

TECHNICAL TERMS

Write a brief definition for the following term. Also, give one synonym that was mentioned in the preceding list of observations.

Unquantifiables ____

____

____

The term unquantifiables refers to those outcomes of solutions that cannot be predicted in numerical terms—at least not as a practical matter. These outcomes, which can be predicted only in qualitative terms, are also called irreducibles.

Aesthetic Quality

The ultimate in defiance of quantification is the visual appeal of a design–its aesthetic quality. This is a criterion that is almost as difficult to verbalize as it is to measure. If you are pressed to express your reaction to a structure, you are likely to respond with "striking" or "neutral" or "ugly" or some other expression of visual impact. The next person, viewing the same scene, might well respond quite differently, and the next still differently. This is natural, since the aesthetic quality of a creation is the manner in which the viewer perceives it; beauty is in the eyes of the beholder, according to an old saying.

YOU DECIDE

What is your reaction to the water tower in Figure 15? Is it ugly, beautiful, or neutral? Explain why.

Figure 15

There is, of course, no way to judge your answer objectively, but see the following paragraph for more on the problem.

Aesthetic quality is the most troublesome kind of criterion; it requires that the designer anticipate the reactions of others, many others, in the face of widely varying individual responses. His final design is what he anticipates will please the most viewers of his creation, and it may or may not coincide with his personal likes and dislikes. Observing what a challenge this is, you can readily surmise why some engineers, architects, and others in the design professions dodge this matter of aesthetics, *and* why authors on the subject of design do likewise. Yet this is an area in which engineers are frequently criticized. So apparently the tendency to slight aesthetic considerations is hurting our creations and our image. As elusive as it may be, this criterion is worthy of your attention, starting now.

Usability

How easy is the oven to operate, the automobile to repair, the production machine to stop in case of emergency, the fire truck to set up for action, the highway signs to follow, the computer to use, the camera to load? These questions call attention to the need for carefully considering the ultimate users of an engineer's creations—the people who must operate or otherwise use them. There are very few (can you think of any?) engineering creations with which humans don't come in contact in some capacity—users, operators, or maintainers—so that a criterion to be considered in almost every problem is usability (some prefer operability).

We have a thick folder of clippings reporting designs that are blunders from the user's point of view. One involves a movie projector with the two switches shown in Figure 16. That seems harmless, except the instruction book emphatically cautions, "DO NOT REVERSE DIRECTION WHILE PROJECTOR IS ON." Surely *that* is going to happen more than once. What makes this example worth mentioning is the remarkably simple way these ON-OFF and reversing functions can be combined into one switch, which eliminates the possibility of reversing with the motor on.

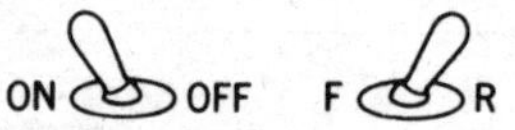

ON OFF F R

Figure 16

YOU DESIGN IT

Sketch a design for a switch that solves the non-reversing-direction-while-the-projector-is-on problem above.

A three-position toggle switch

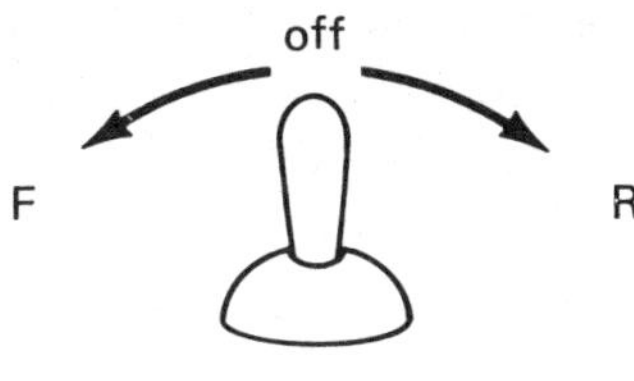

Figure 17

Some other examples of poor design are: an automobile with the steering column placed so that the driver's foot is easily blocked as he tries to shift it from accelerator to brake; the highway interchange that seems to invite drivers to enter by an exit; truck cabs outrageously awkward to get in and out; hazardous appliance plugs; and lecture hall controls in ridiculous locations. Surely you have encountered similar cases; everyone seems to have their pet gripes in this respect. Who is to blame? Engineers, architects, industrial designers, and automobile stylists who simply haven't given enough attention to the people who ultimately will use—or try to use—their creations.

In a majority of instances designers have done their job well in this respect, and sometimes exceptionally well, as illustrated by Figure 18.

Figure 18. To score well with respect to the criterion usability, a machine like this grader must provide for good operator visibility, controls that are easy to learn and readily accessible, a seating arrangement that minimizes operator fatigue, and a variety of other operator-oriented features. And this one does; the designer has done an excellent job in these respects.

Sometime, if you have the opportunity, take a look at the shape-coded controls and other considerations of the user in the cockpit of a commerical airliner. (A lot of attention has been given to the designing of instruments and controls to prevent pilot goofs.) And we notice that the designers of a procket calculator are thoughtful of users. For instance, the tilt of the display improves visibility; the click of the function keys assures me that the function has registered; and the ON-OFF button is located to minimize the chances of accidentally switching the machine on when sliding it into a case or pocket. How do the designers of graders, aircraft, calculators, and a majority of other engineering contrivances avoid blundering in their consideration of the user?

THINK IT THROUGH

Stop for a moment and think about that last question. How do designers go about avoiding blunders that will affect of the user of their products?

____

____

____

A few of the many possible answers to this exercise appear in the following paragraph.

For one thing, they devote a lot of forethought to the user—what he needs, his habits, and his physical, sensory, and mental limitations. Furthermore, they frequently conduct field trials with prototypes to evaluate usability. In addition, they request help when they need expert advice in such matters, which they can get from handbooks and specialists in the field of *human engineering* (some prefer the term *human factors*). Finally, they apply a lot of common sense and thereby avoid the design blunders you read about in *Consumer Reports*.

The tendency of some practitioners in the design professions to slight the criterion of usability is puzzling to us, partly because so much of this is common sense, and partly because such considerations are the most interesting part of our work as engineers.

Safety

Certainly, of the alternative schemes considered in the design of an improved tire mounter, some were more likely to injure workers than others, and this does matter. Although the designer did not attempt to measure this characteristic of alternatives, this in no way reflects the importance he attached to the safety criterion. The fact is, safety of proposed solutions is virtually unquantifiable. As elusive as it is, however, this is a criterion engineers cannot afford to slight. A mass of consumer protection legislation and a dramatic increase in liability for product-caused mishaps make thorough consideration of safety a must.

THINK IT THROUGH

We are about to mention a number of things an engineer must do to fulfill professional obligations in regard to safety. You should be able to anticipate a few. What do you think some of these obligations are?

____

____

____

Some answers to this exercise appear in the following paragraph.

In order to fulfill his professional obligations with respect to safety, an engineer must:

- Avoid design blunders, such as miscalculation of the shear force on a strategic load-bearing pin in a piece of agricultural machinery, causing it to fail in use.

- Anticipate how his creations will be used in the field, not simply how they should be used. (Manufacturers are liable for product-caused accidents even when the injuries arise from misuses of the product.)

- Strive to foresee all of the possibilities for injury to the user—even those that might result from the user's carelessness. Try to accident-proof the creations accordingly (for example, through hand guards, electrical interlocks, overload sensors, and emergency switches).

- Exhaustively test his solutions and strategic components thereof, in the laboratory and in the field, for vulnerability to hazardous failures (for example, an axle prone to break because of metal fatigue).

- Remain informed about the numerous codes and standards prescribed by law, professional and industrial organizations, Underwriters' Laboratories, and the like.

This is not all that safety consciousness implies for the engineer, but this sampling suffices to demonstrate that there is a lot to this matter.

Product safety considerations present a special challenge, and not only because of their importance. Some of this challenge derives from the quantification difficulty. Furthermore, this matter of foreseeing all the accident-causing eventualities in the manufacture, use, and servicing of a product certainly doesn't make things simple either. Take this case.

An automobile company was sued (and subsequently paid) because a front brake hose burst, causing an accident. It burst because a front wheel had been rubbing it on turns, and this rubbing was traced to sloppy assembly and inspection at the factory.

THINK IT THROUGH

What can an engineer do about such an assembly and inspection problem as the rubbing brake hose?

____

____

____

____

An answer to this exercise appears in the following paragraph.

The prevailing attitude of the courts today is that this rubbing and subsequent failure because of improper assembly and careless inspection is foreseeable, so it behooves the engineer to anticipate such unlikely chains of events and, if possible, foolproof his designs accordingly. Reading the case histories of product liability suits is a journey into the bizarre, and although you might question the foreseeability of many of the accidents, the courts don't.

TECHNICAL TERMS

Write a brief definition for each term.

1. Aesthetic qualities ____

____

2. Usability ____

____

____

3. Human engineering—include one synonym ____

____

____

1. Aesthetic qualities refers to such things as visual appeal as to form, color, and composition; and to such things as the pleasing sound produced by a product, or the feel of its operation. These qualities are generally not quantifiable.

2. Usability refers to the closeness with which an engineer's solution is adapted to the people who use and service it.

3. Human engineering is a specialty at the interface between psychology and engineering, concerned with usability of engineering creations. So human engineers are especially interested in ease of operation, safety, serviceability, ease of learning, and other human-oriented criteria.

YOU DESIGN IT

List the criteria you believe should be considered by designers of one of the following: a highway, an artificial hand, a method of assembling automobiles, or the automobile itself.

1. Criteria for the highway ____________________

____

____

____

2. Criteria for the artifical hand ____________________

____

____

____

3. Criteria for the automobile assembly method ____________________

____

____

____

4. Criteria for the design of the automobile itself ____________________

____

____

____

1. Criteria for the highway: right-of-way cost, design and construction cost, social disruption, personal hardship, noise, pollution, aesthetic appeal, safety, maintenance cost, travel benefits, cost to users, and expandability.

2. Criteria for design of an artificial hand: cost to develop, cost to manufacturer, func-

tions performed and effectiveness of them, ease of operation, appearance, reliability, operating cost, skill required, and installation ease.

3. Criteria for automobile assembly method: development and construction cost, cost-of-operation, reliability, safety, environmental effects, skills required, physical and psychological demands, energy required, expandability, adaptability to changes in vehicle design, and adaptability to changes in production rate.

4. Criteria for the design of an automobile: development cost, manufacturing cost, visual appeal, efficiency, noise level, safety, maintenance cost, riding comfort, steerability, space, expected life, energy consumed, materials consumed, reliability, ease of operation, and disposability.

SOME EFFECTS OF INCREASING COMPLEXITY

Some of the criteria that come up repeatedly are increasing in importance as engineering creations become more complex. One of these is *reliability* which, by the way, has a very specific meaning to an engineer. To him, reliability is the probability that the component or system in question will not fail during a specified period under prescribed conditions. For example, the reliability of video tubes made by the Zeon Company is 0.95 for the first year of operation under "normal" usage, vibration, and temperature conditions. Reliability is especially important when failure is costly, which is true of an amplifier in a transatlantic cable under 3000 meters of water or in an orbiting communications satellite. It is also important when people are highly dependent on the system, which is true of public water supply systems, defense missiles, and elevators in a 40-story building.

Reliability, along with criteria such as repairability, maintainability, availability, and usability, are more important not only as a result of complexity; the heavy investments required by many modern engineering creations and our dependence on them also add weight to these criteria. A whole production system depends on that simple little tire mounter; a corporation can "go to pieces" if its computer breaks down for long; and we all know what happens when a city's electric power system cuts out for a few hours!

Note that specific criteria like reliability and repairability and, ultimately, the total cost of a solution, can be improved through simplification. Simpler mechanisms, simpler circuits, simpler functioning, simpler shapes, and so forth are generally worth striving for. This fact, combined with seemingly irreversible trend toward greater complexity of man's creations, prompts this special plea on behalf of the simplicity criterion.

A PLEA FOR SIMPLICITY IN DESIGN

Usually, among the many solutions to a problem, some are outrageously complicated; others are refreshingly simple but no less effective than their complex counterparts. The tire mounter provides an excellent example. The previous machine was unnecessarily complicated; it had 60 moving parts and required electricity and compressed air. The new device is a sharp contrast: it has only a few moving parts, requires electricity only, and there is practically nothing to fail or maintain. This comes from good engineering.

Another example: an artificial satellite tends to tumble indefinitely as it orbits Earth, yet for certain applications one side of the satellite must remain pointed toward Earth at all times. This is true of the weather satellite; its cameras will hardly do us any good if they are pointing out into space. Therefore, such a satellite must have an orientation system that prevents it from tumbling and keeps it properly aimed. A long boom attached to the satellite provides a beautifully simple solution to this imposing problem, as explained by Figure 19.

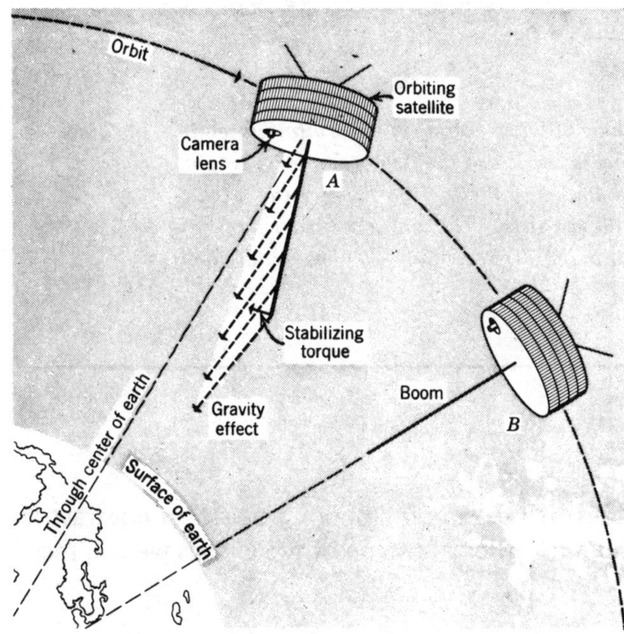

Figure 19. The pull of gravity on the end of the boom closest to the Earth is greatest at that point (the force of gravity varies with the square of the distance to the Earth's center). This swings that end toward an imaginary axis extending from the satellite to the Earth's center. This gravity-induced torque causes the structure to seek the desired vertical orientation. In position A, gravity is in the process of orienting the still-tumbling satellite. Eventually the structure reaches and remains in orientation B. In space jargon this is referred to as the gravity-gradient method of satellite stabilization. (Incidentally, since there is no air resistance, this structure will act like a pendulum on a pivot and swing to and fro indefinitely unless some provision is made to dampen this oscillation. This is done, for example, by attaching a tiny weight to the tip of the boom by means of a coil spring, so that the weight flops to and fro out of phase with the boom and eventually kills the oscillations. It is possible that by chance the satellite-boom structure will orient itself with the satellite facing the wrong way, which means that the satellite's cameras would be forever aimed at outer space. In this event, the boom is retracted, the satellite is allowed to tumble, then the boom is reextended, and this process is repeated until by chance the structure assumes the desired orientation.)

To fully appreciate the virtues of this solution, you must know something about the alternatives. You would be amazed at the complicated systems that have been used or have been proposed to achieve this same purpose. One system uses electronic horizon sensors in conjunction with a system of gas jets, which means that it requires compressed gas and electricity (the gravity gradient method requires no power). In addition, it has about 40 times as many parts, weighs about four times as much, costs about 50 times more, takes more space, and is considerably less reliable than the gravity-gradient system. These are no small matters.

The stabilizing boom on a satellite may be up to several hundred meters long, and it obviously cannot be protruding from the satellite during the launch phase. So how does a satellite "grow a boom," say 25 meters long, after it is in orbit? Many elaborate schemes were considered, but the solution eventually conceived is remarkably simple (see Figure 20), thanks to an engineer's conviction that there must be a better way.

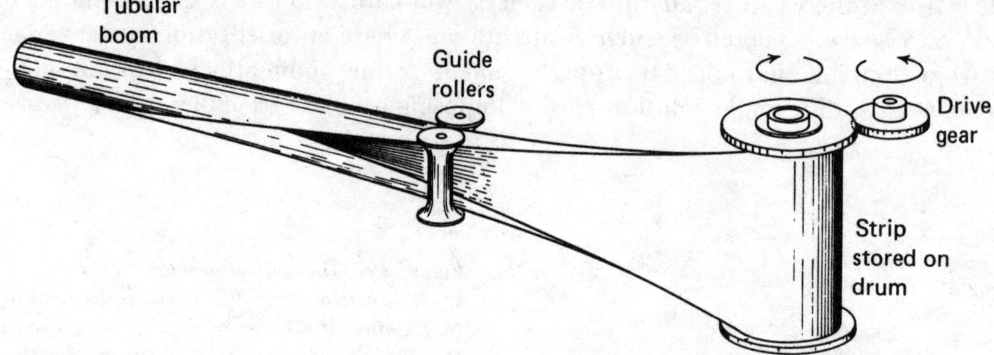

Figure 20. Design concept for "growing" a long tubular boom on an orbiting satellite. A flat metal strip the length of the desired boom is treated so that is normal form is tubular. Then it is rolled up flat onto a spool and stored in the satellite. After the satellite is in orbit, a signal actuates a motor that unrolls the strip. As it unfurls, the strip returns to its tubular form and, when fully extended, is the desired boom. Tubes hundreds of meters long can be so constructed.

THINK IT THROUGH

We are about to mention some of the virtues of this simple "boom grower" in contrast to more complex solutions. You should be able to anticipate some of the virtues we will mention. Try.

____

____

____

____

Some answers to this exercise appear in the following paragraph.

What are some of the virtues of this simple "boom-grower," especially in contrast to the solution that might have evolved? The former is compact, light, reliable, relatively inexpensive, and requires very little power. In fact, the spring action of the metal provides the energy needed, so that it can be operated without a motor. Engineers call this an *elegant solution*, because it achieves the desired objective in an uncomplicated manner.

TECHNICAL TERMS

Write a brief definition for each term.

1. Reliability ____

____

2. Simplicity ____

____

3. Elegant solution ____

____

1. Reliability refers to the probability that a designed solution will not fail under prescribed conditions during a specified period.

2. Simplicity in design refers to the fact that such things as the number of moving parts, the complexity of component interrelationships the number of inputting sources of energy or direction, and so on are all reduced to the absolute minimum necessary to do the job.

3. The term elegant solution refers to a solution that achieves the desired objective in a cleverly uncomplicated manner.

An engineer should not be content until he has simplified to the maximum possible extent the mechanisms, circuits, method of operation, maintenance procedures, and other features of solutions. Ordinarily there is a big difference between a workable solution at the time it is conceived and that solution after it has been effectively simplified.

The simplicity of the adopted solutions in the three cases above is impressive. However, simple designs, per se, are valuable only in the satisfaction (don't underrate it) that they bring to their creators. The payoff comes from the economy of manufacture, low maintenance cost, and high reliability that results from simplicity.

Observe how a cleverly simplified solution tends to deceive the untrained eve. It belies the difficulty of the problem and all the knowledge, skill, and effort that went into it. This is certainly true of the examples just cited. How easy it is to grossly underestimate what underlies this simplicity, especially if you are unaware of the overly complex solutions that could have evolved (and sometimes do). When an engineer's solution has reached this deceptively simple state, he can consider his job done well.

A mark of an exceptionally creative person in almost any field is the simplicity of his or her creations. Observe the few lines required by a good cartoonist to create the desired effect, the remarkably few brush strokes evident on a good watercolor, the few words required by the skillful author, the simple lines that make the great works of architecture, or the striking simplicity of the three engineering creations just described. Yes, there is art in engineering and, by the simplicity criterion, these solutions are exceptionally artistic.

Could anyone have thought of the solutions described for the tire-mounting, boom-growing, and satellite stabilization problems? If not, what special knowledge was required? A clever person without a technical education but with mechanical inclinations could have thought of the solution concept underlying the new tire-mounter, but he would have been hard put to refine his concept and to specify the details of a workable device. In the case of the boom-grower, a good machinist could have come up with the concept, but the engineer who developed a functioning mechanism had to have special knowledge of metals, heat treating, stresses, mechanisms, and cantilever beams. Only a person familiar with the principles of mechanics, gravitational phenomena, oscillatory systems, and the like would be equipped to conceive, let alone develop the gravity-gradient idea.

The nub of the solution in each of these instances is pure invention, the product of an engineer's ingenuity. But without special technical knowledge, it would have been virtually impossible to convert the basic idea to a workable solution in any of these cases. So although invention is a necessary and very important part of engineering, it is hardly sufficient.

This analysis highlights a major difference between present engineers and engineers of preceding periods like Edison, Whitney, and Watt. All are inventors, but there is a substantial difference in the amounts and types of knowledge available to engineers of the past and those of the present.

YOU DESIGN IT

Assume that the following memorandum is directed to you. Thus it is your job to prepare the preliminary report that Chief Engineer Hogan requests. The primary purpose of this assignment is to give you an opportunity to practice what you have read about design techniques. As a minimum, we suggest you do the following:

1. Define this problem. The form supplied below is recommended.

2. Conceive as many solutions as you can in, say, 15 minutes of intensive creative thought. Hopefully, you will produce an impressive alternatives tree for each major solution variable.

3. Evaluate with respect to relevant criteria three partial solutions for one or more solution variables.

Solution variable ________________________________

Brief evaluation of partial solution A: ________________________________

Brief evaluation of partial solution B: ____________________

Brief evaluation of partial solution C: ____________________

4. On the basis of evaluations like the above, you might like to prepare a sketch of your preferred solution, which will be a composite of the best partial solutions (that is, the best support system, best skin material, and so on).

DESIGNWELL ASSOCIATES

Consulting Engineers Inter Office Memo

TO: Stu Dent, Project Engineer

FROM: James C. Hogan, Chief Engineer

This is to verify our conversation of yesterday. The Red Cross seeks to develop a low-cost disaster-victim shelter. They have asked us to submit preliminary plans for a system that meets their specifications.

They have in mind an air-transportable shelter weighing not more than 12 kilograms, occupying no more than one-tenth of a cubic meter unassembled, and suitable for 4 persons. They want these structures to be readily interconnectable to accomodate larger social units.

Remember, it will be assembled by unskilled persons, many of whom cannot read.

The Red Cross plans to stock 500,000 of these units, distributed over a half-dozen sites around the world. For fast deployment, these units should be stored for quick loading and unloading.

I am asking you to prepare a preliminary design and to get your report to me in the very near future.

PROBLEM DEFINITION

Problem Formulation

State A in General Terms

State B in General Terms

Problem Analysis

Restrictions:

Criteria:

Solution Variables:

Volume:

Usage:

FORMULATION

DEFINITION OF "SHELTER" PROBLEM

Unsheltered Victim ⟶ Sheltered Victim

ANALYSIS

Restrictions:

- Must be air transportable.
- Cannot weigh more than 12 kilograms.
- Cannot occupy more than 0.1 cubic meter unassembled.
- Must be assembled by unskilled labor without the benefit of printed instructions.
- Must accommodate 4 persons.
- Must provide for interconnection with adjacent units.

Criteria:

- Cost
- Ease of assembly
- Weight
- Space consumed
- Livability
- Comfort
- Vulnerability to damage
- Safety to accupants
- Disposability

Solution variables:

- Skin material
- Method of support
- Method of assembly
- Method of storing in transit
- Basic shape
- Specific dimensions

Volume: 500,000

Usage: (Depends)

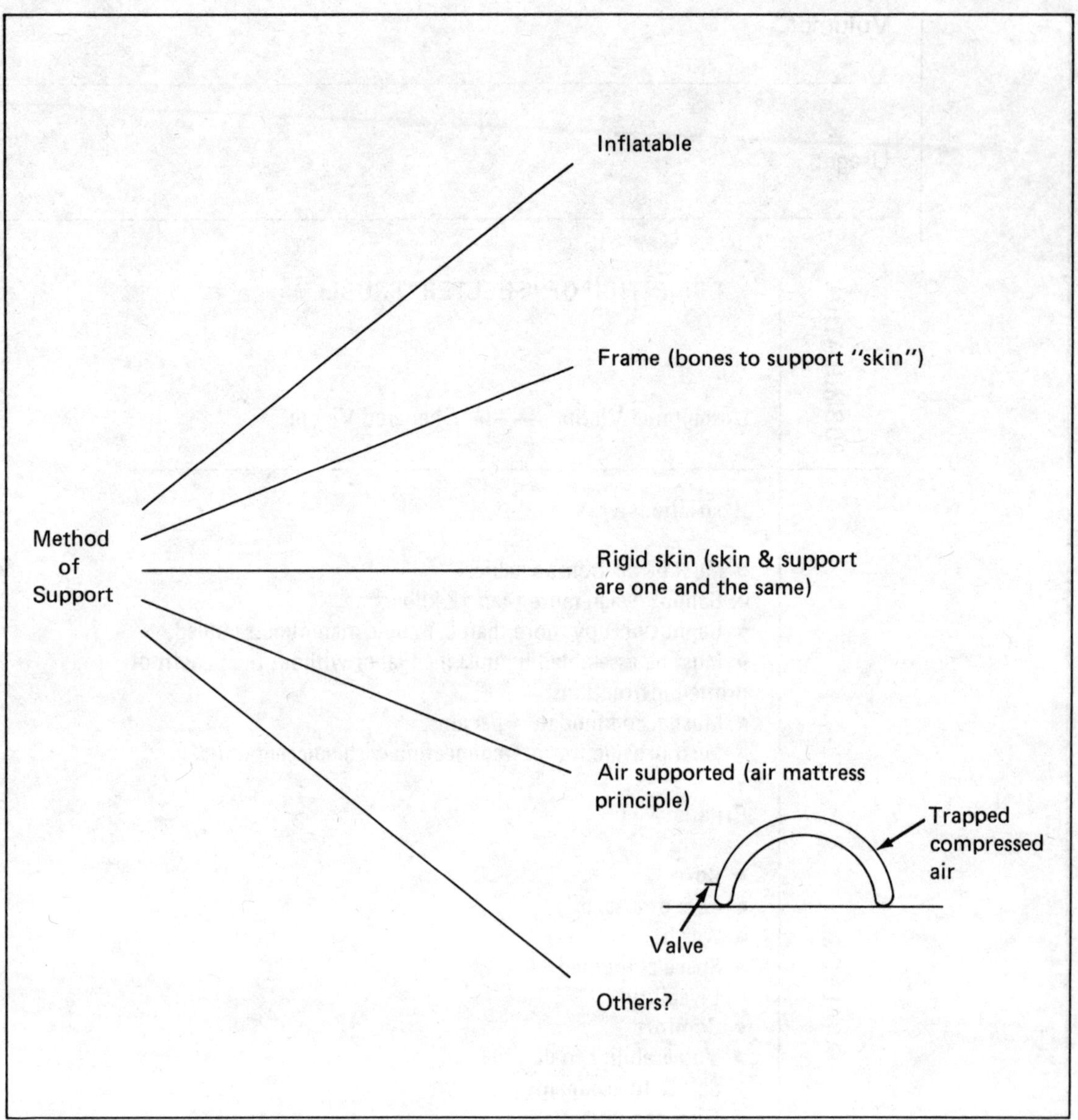

Figure 22. A sample alternatives tree showing some alternative partial solutions for one solution variable in the disaster-victim shelter.

SELF QUIZ

1. What four steps must be followed in most decision-making processes?

2. Define the term *cost-benefit ratio*.

3. Why is it important for the designer to predict the performance of all alternative solutions in the same units?

4. Define the term *unquantifiables*. Give at least one synonym. Give a few examples of unquantifiables.

5. Define the term *human engineering*. Give one synonym for this term.

6. List three ways that engineers try to avoid blunders in their consideration of the user of their creations.

7. List three ways that engineers try to fulfill their professional obligations for safety.

1. The four steps to be followed in most decision-making processes are:

- Select and weight criteria.
- Predict performances of alternative solutions.
- Compare predicted performances.
- Make the design decision.

2. The benefit-cost ratio, also called the cost-benefit ratio or effectiveness-cost ratio, is the relationship between the total expected costs involved in creating, implementing and utilizing a solution and the total expected benefits for the users and for society.

3. It is impossible, or at least not practicable, to compare predictions of the performance for alternative solutions unless these performances are measured in the same units. Of course, the most convenient unit for measuring performance is often monetary costs and benefits.

4. Unquantifiables, or irreducibles, are criteria that cannot be measured in numerical terms. Aesthetic qualities, safety, usability, and moral value are a few such criteria. Such criteria must be expressed and evaluated in qualitative terms.

5. Human engineering, or human factors, refers to the activities of designers that attempt to anticipate fully the needs of users of their creations so that the product will be as usable as is humanly possible to make it.

6. Engineers and designers attempt to avoid blunders in considering the users of their creations when they:

- study users' needs
- study users' wants
- study users' habits
- study users' physical, sensory, and mental limits
- conduct field trials with prototypes
- consult a great variety of reports, manuals, and handbooks
- consult with specialists in human engineering or human factors
- retain common sense in the midst of technical complexities

7. Engineers attempt to fulfill their obligations to safety when they:

- take great care not to miscalculate the performance of materials and methods to be used in their designs
- anticipate how their creations will be used in actuality, and are not lulled by a sense of how they should be used in theory
- build in designs to prevent accidents, even with misuse
- thoroughly test prototypes and production models, both in the laboratory and in the field, to discover unexpected weaknesses
- continue to study and learn all the developments in their fields that affect safety, as well as other professional concerns

Learning Segment 6
DEVELOPING SPECIFICATIONS AND IMPLEMENTING AND COMPLETING THE DESIGN CYCLE

LEARNING OBJECTIVES

- Understand and perform the specification phase of the design process.

- Understand the importance of events after the design process has been completed; these include implementation events, follow-up events, and reactivation events.

- Recognize that the design process is a basic problem-solving process that can be applied to both technical and nontechnical problems when appropriate.

INTRODUCTION

At the conclusion of the decision phase, the preferred alternative is in the form of a solution concept (some call it a design concept). This is the major features and operating principles of the solution but not such details as particular materials and exact dimensions, which are still lacking. Thus the development of a solution concept is not the end of an engineer's encounter with a problem. A solution in the form of pages of rough sketches, notes, and computations, with many details the designer has yet to put on paper, is unsatisfactory for several reasons.

One reason, and few engineers can escape it, is that they do not have the final word on implementation of their solutions. Almost invariably there are others above them whose approval they must gain. Although acceptance was virtually automatic, the designer of the tire-mounter had to get a green light from his immediate supervisor, who is the manager of engineering, and from the superintendent of manufacturing. Failure to win the backing of either would have killed his proposal.

And let's face it, the people who can veto an engineer's works are usually hard-nosed businessmen or tightly budgeted public officials who must be convinced. To be convincing, proposals must be quite specific about the nature of the proposed solutions, as well as about their virtues and limitations. Consequently, the designer must describe in considerable detail the physical attributes and performance characteristics of what he has in mind.

There is another and very practical reason for this. An engineer who designs a machine or structure never actually builds, operates, or services his creation. This makes it essential that he amply document and effectively communicate the proposal to the persons charged with construction, operation, and servicing, so that they can satisfactorily fulfill their responsibilities.

SPECIFICATION PHASE OF THE DESIGN PROCESS

The upshot of all this is the specification phase of the design process which, among other things, ordinarily yields detailed, dimensional drawings of the final solution. These are usually referred to as engineering drawings (sometimes as the prints) and, for a complex system like a bridge or chemical plant, they may number in the thousands.

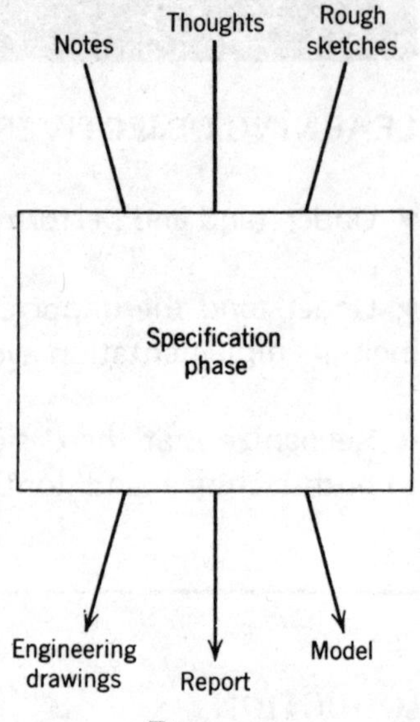

Figure 23

Another medium for communication of the designer's solution is the engineering report, usually a rather formal and voluminous document describing and evaluating his proposal, mainly in words. Many a fledgling engineer finally comes to appreciate the importance of verbal skills when preparing his first report of this kind.

Occasionally the prints and report are supplemented with a model, which may be a working model for testing and demonstration purposes. And, finally, the engineer's communication efforts may also include an oral presentation of the proposal to superiors, clients, and sometimes the public. That, too, calls for skills worth cultivating.

THINK IT THROUGH

What sort of work will the specification phase of the design process require of the engineer? What sorts of things will have to be specified?

____

____

____

____

____

____

____

Some answers for this exercise appear in the following paragraph.

The specification phase is likely to involve a lot of detail work. Draftsmen and other technical assistants take over some of this burden but, in general, the engineer must specify the types and properties of materials to be used, exact dimensions and tolerances, methods of fastening, operating temperatures, voltages, and so on. For obvious reasons, this seldom proves to be the most enjoyable part of the work.

THE DESIGN CYCLE

The delivery of a final report terminates the engineer's design efforts on a given project but certainly not his responsibilities. The professional extends himself to gain acceptance of the design, to oversee its installation and use, and to accomplish other postspecification functions, identified in Figure 24. This cycle of events is worth examining.

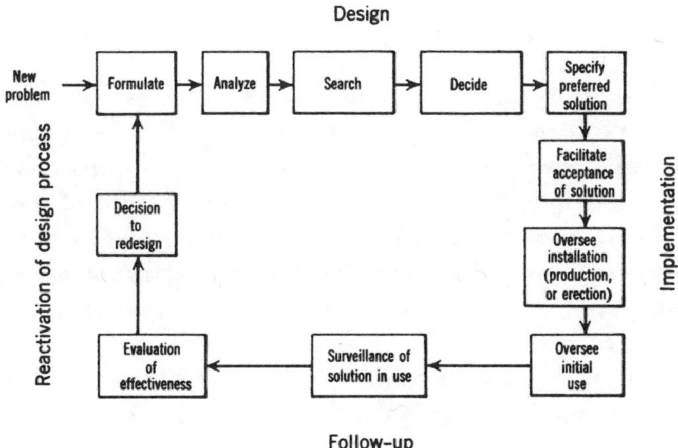

Figure 24. The design cycle.

Implementing a Solution

Never assume that a solution will automatically be adopted, properly constructed, and used as intended. Many things can go wrong, and measures must be taken to prevent them from doing so between the time a solution is proposed and the time it is an accomplished fact.

For example, measures are needed to ensure that a solution will be adopted by the appropriate people. Engineers often begin their careers with the mistaken impression that, if their proposals are technically and economically superior, they will naturally be implemented. But engineering is ordinarily a staff function in an organization, so that engineers issue recommendations, not commands. This, plus the fact that differences of opinion do arise, makes it imperative that careful attention be given to this matter of gaining solution acceptance.

Young engineers are inclined to become discouraged if several of their proposals have been rejected. They are inclined to blame other people, the organization, anything or anyone but themselves. But the chances are that they have underestimated the need for effective presentation of their proposals, for convincing others on the worth of their ideas, for a certain amount of realistic compromise, and for careful planning to minimize resistance to change.

Follow-Up

Periodic monitoring of solutions in use is especially valuable as a means of improving future designs. Rare is the engineer who cannot benefit by watching people operate the computer terminal he designed, service his automobile engine, or follow his highway directional signs. Not only is this kind of activity educational, but it is likely to be one of the more interesting aspects of his work.

Reactivating the Design Process

Periodic evaluation of solutions in use also provides a basis for deciding when to redesign. No solution to a practical problem remains superior indefinitely. Better methods are discovered, new demands arise, new knowledge accumulates, conditions change, and physical depreciation occurs. Consequently, a point is reached in the life of a design at which it is profitable to seek a better solution. An engineering department can intelligently decide when to engage in redesign only if the current solutions to problems within its realm are periodically appraised.

The design cycle is complete when, after a solution to a problem has been devised and used over a period of years, it appears that redesign will prove profitable, and the process of designing a superior solution is again initiated.

APPLICABILITY OF THE DESIGN PROCESS

What you have been reading about in this unit can be applied to a broad range of technical and nontechnical problems.

Technical Applications

You have already had a look at the "boom-grower" satellite. Here is some insight into the birth of that mechanism: An engineer was asked to find a means of constructing a 25-meter boom, 5 centimeters in diameter, solid or tubular, on a 100 × 60 centimeters hatbox satellite, *after* it is in orbit. Figure 25 indicates what he did first: he set up the problem.

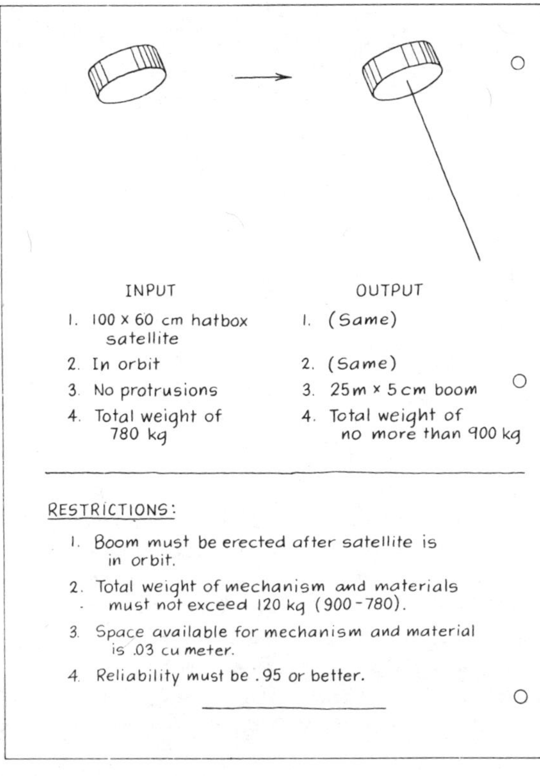

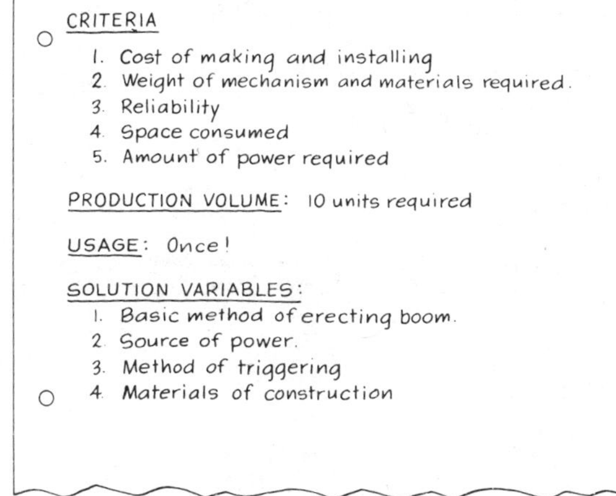

Figure 25. These two pages from the engineer's notebook indicate how he defined his problem. (We have eliminated some of the details and simplified others.).

MEMORY JOGGER

What two steps in the design process have been performed in the notebook pages shown above?

Formulation of the problem, analysis of the problem

When he was satisfied with his definition, he went to work on alternative solutions. Figure 26 shows some of the possibilities he came up with. To him, in restrospect, and probably to you now, Figure 26 makes the whole thing look simple and obvious. But what you see in this figure is the product of about three days (not all at one stretch) of thinking, talking, reading, visiting, corresponding, telephoning, sketching, daydreaming, sweating, scratching, and smoking.

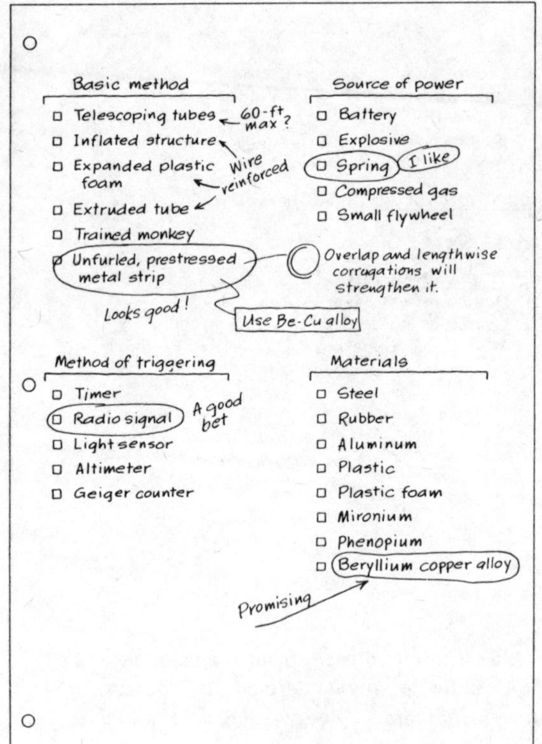

Figure 26. Another page from his notebook (again, we have done some simplifying), showing some of the alternative partial solutions that he accumulated for the major solution variables. Based on what we have said, you are not surprised that we are happy with the manner in which the used solution variables to guide his search. The thoroughness with which he capitalized on this approach may well account for the elegance of his final solution.

MEMORY JOGGER

This engineer has now formulated the problem, analyzed the problem and searched for alternative solutions. What steps of the design process must still be performed?

He must decide and specify his preferred solution.

What followed was the most time-consuming part of the task: gathering performance data and cost figures, adding details and more details, calculating, experimenting, comparing alternatives, and so on, until he had tentatively decided on the mechanism pictured earlier. A working model of that concept was constructed and tested, after which minor changes were made. He continued this test-and-modify cycle until he was satisfied. The rest of the job was a matter of preparing detailed, dimensional drawings and his final report.

Those last few sentences make the task of converting concept to reality sound so simple and routine; some of it, yes, but much of it—no.

Here is a brief glimpse at the kinds of things he ran into. The extended boom is subject to several types of forces, the magnitudes of which must be predicted. One of them is a bending action generated by sunrays striking one side of the boom (something that many of us could easily have overlooked). The challenge here is to predict the heat-generated forces in the boom, then to do the same for other forces, and then to determine the structural shape and material thickness necessary to withstand the resultant of all forces. This is only one of hundreds of details that must be resolved to convert the original design concept to a fully specified, workable solution, details the layman quite naturally overlooks as he marvels at the remarkably simple result.

The preceding case is a beautiful application of the design process to a technical problem. *The same basic process applies regardless of the nature of the problem.* To illustrate this broad applicability, here is an elaboration on the spectrum of activities bounded by *development engineering* and *sales engineering*.

Task A, illustrating development engineering, involves design of a device that will directly convert the spoken word to printed form. The input is an oral message, the output must be a record of that message on paper.

Task B, representative of sales engineering, is at a company that manufactures electrical equipment (motors, transformers, and so on) and assembles these components into power systems designed to suit the unique needs of individual customers. Most of these customers are factories, refineries, printing plants, and the like. Task B involves calling on a prospective customer, in this case a paper-manufacturer who plans to construct a new plant. At the potential customer's invitation, the engineer familiarizes himself with the paper-making process, makes a thorough investigation of the company's needs, and then designs a power system adapted to the customer's process. In doing so, he relies heavily on components manufactured by his employer. He submits his solution along with the price to the potential customer. If his system is purchased, he will oversee its installation and remain in contact with the job until it is running smoothly.

THINK IT THROUGH

What is the basic difference between the task assigned to the development engineer above and the task assigned to the sales engineer?

____

____

____

____

____

An answer to this exercise appears in the following paragraphs.

Task A is concerned primarily with the *conception* of a new device. There is a minimum of past experience on which to rely; much original thought is required; some research may be necessary; the work can get steep technically.

Task B is mainly the *application* of existing devices and components in the satisfaction of the unique needs of specific customers. For this type of work the engineer can rely on a large backlog of experience in designing such systems in similar situations. Although each situation he encounters is different to a degree, the work involved is hardly what you would consider pioneering. Whereas the main challenge of Task A stems from the original nature of the work, the main challenge of Task B is in acquiring a thorough understanding of the needs of the customer.

These two tasks lie close to the extremes of a wide range of types of problem-solving activity. Along this spectrum lies such tasks as the design of an interplanetary transport, artifical heart, automobile, cement plant, bridge, computer, pipeline, or undersea vessel. In all cases there is a problem to be defined in terms of a transformation, criteria, restrictions, and so forth; there are alternative solutions to be identified, choices to be made, a proposal to be specified.

Nontechnical Applications

Picture yourself with a set of simultaneous equations for which you must achieve a transformation. In order to do this, you should not automatically resort to a particular method. There are alternative solutions, such as the substitution method, graphical means, determinants, or computer.

MEMORY JOGGER

Does the term *solutions* sound strange to you in the context in which it has just been used? It shouldn't if you recall the definition for the term *solution*. What is that definition?

____

____

The answer to this exercise appears in the following paragraph.

If the use of the word *solutions* here bothers you, recall that solution was defined as "a method of achieving the desired transformation." This may well run contrary to your customary use of "the solution," as well as of "the answer."

These solutions, or methods, differ with respect to criteria like time required, equipment needed, and accuracy attainable. Naturally, you will apply these in selecting the best method for the particular situation at hand. Furthermore, a teacher would have imposed at least one restriction—that you solve it. Thus, this and other mathematical problems can be structured and approached according to the pattern introduced in this unit, with some benefit to you if you are willing to give the proper emphasis to alternatives and criteria.

The transformation involved in a physician's typical problem is obvious enough. He works under restrictions set by statute, professional code, and sometimes the patient. He has alternative solutions available in almost every case; for instance, for ulcers, the main alternatives are diet, surgery, radiation, and medication. Certainly he applies a number of criteria in selecting the best treatment. Among them are probability of success, time required, cost, and discomfort.

A learning program such as the one you are now studying is intended to transform certain qualities of students. Suppose that you have been assigned to develop an introductory learning program in engineering. Surely, in defining this problem, you would thoroughly specify the knowledge, skills, and attitudes you would want to find in students when they had finished. Similarly, for the students' "input," you would want to know something about their backgrounds, misconceptions, apprehensions, and other facts. With this specific information on "input" and "output," you would know the transformation that your course must bring about.

Ordinarily, restrictions on such a learning program are imposed by available space in the book, the nature of the printed page, and the like. In this instance, you are told that it must be less than 400 pages. Among the criteria you would employ are benefit to the student, appeal to her or him, student hours required, resources required, and development cost. In this problem, too, there are solution variables, such as method of communicating to the student and means of obtaining responses and giving feedback. For each of these there are numerous alternatives.

Since mathematical problems, medical problems, educational problems, business problems, you-name-it problems all involve transformation, restrictions, criteria, and so forth, they are *basically* no different from what, up to this point, have been called technical problems. It is conceivable then that you have learned something about problem solving that you can apply personally, as well as professionally.

Now don't get us wrong; we are not suggesting that you formally structure every problem that you come upon along these lines, or that you always engage in an intensive search for solutions. We don't recommend this; we don't do it ourselves.

THE DESIGN PROCESS—AN OVERVIEW

Clearly we are discussing the principal activity in engineering—problem solving. However, in the conversation and literature of engineers, this activity is often referred to as design, or engineering design, or the process of design, or the design process. Such terminology is not standard and, furthermore, interpretations of a given term vary among users. In view of this, let there be no doubt about the preferred term in this book—*design process*.

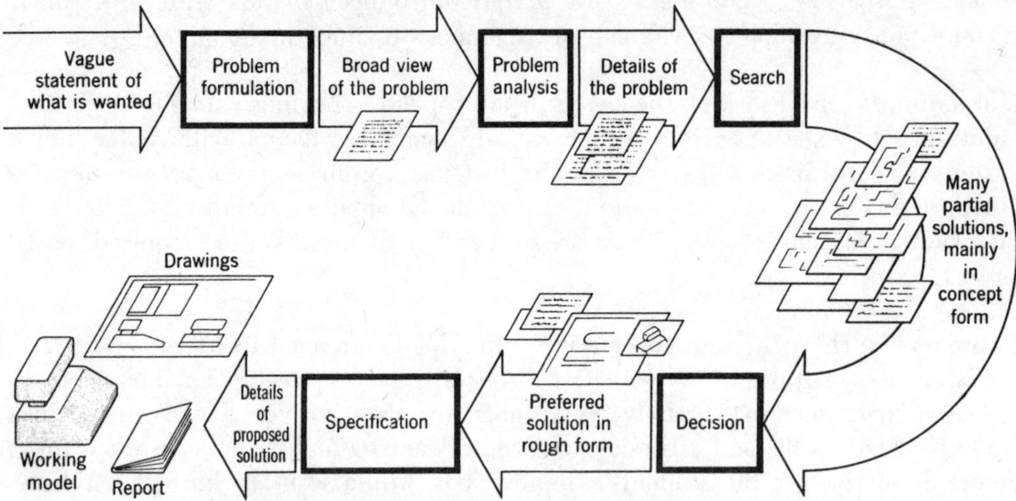

Figure 27. The design process, picturing the input and output of each phase.

As used herein, the design process is what this unit is all about. It begins with a vague statement of objective, ends with the specification of a solution, and embraces all activities and events between. As pictured in Figure 27 this process is a series of stages in the evolution of a solution. The purpose of each phase is different; the type of problem-solving activity that predominates in each is different. However, these stages have no sharply defined boundaries, nor are they the orderly progression of distinct steps the idealist would hope for. There is a certain fuzziness as the emphasis shifts from one phase to the next. Occasionally the designer will come up with solutions while he is defining his problem or, during the search phase, he may decide to reformulate the problem. Similarly, it is impossible not to do some evaluating during the search phase. Chance plays a significant role in this process; new information and new ideas are uncovered unpredictably; blind alleys are encountered; occasional backtracking is inevitable. The process is clearly and unavoidably iterative.

PRACTICING WHAT YOU HAVE BEEN READING

With the temptations, pressures, and pitfalls that face you on the job, it behooves you to be prepared to perfect your design skills. One consequence of being unprepared in this respect is that you surely will be victimized by certain habits of thought that hurt your performance—habits such as giving in it the temptation to bypass problem definition. If you are to overcome such habits, you must work at it. That means reading and hearing about how to do it and, since this is a skill, it also means practice. While practicing, evaluate your problem-solving approach and discipline your mind accordingly. Such efforts will pay off in superior design performance over the years.

CHECK IT OUT

Think about your design (problem-solving) skills. Are you strong in some areas? Are you weak in some areas? What should you do to capitalize on your strong areas and strengthen your weak areas?

____

____

____

____

A PRECAUTION ON THE TERMINOLOGY OF PROBLEM-SOLVING APPROACHES

Of late there has been a flood of books and articles proclaiming the virtues of a "new" problem-solving approach. As a consequence, sociologists, psychologists, business executives, economists, hospital administrators, government people, and all readers of the help wanted ads in the big city papers are talking about *systems analysis*. However quick writers may be to use the term and speak of its virtues, they are just as slow to tell us what it is. As it turns out, there are numerous interpretations. Some authors define systems analysis literally. To them, it is simply the process of identifying subsystems, components, and their interrelationships in order to better understand a given system. Some writers, however, obviously use the term to means systems *synthesis*. Others feel it relates only to computer systems, which is nonsense. Many use it in a sense that puzzles an engineer, since they seem to mean what he has known for a long time as the design process. When portraying systems analysis, its enthusiasts frequently speak of a "total system approach" to a problem (meaning a broad definition), of an intensive search for alternatives, of benefit-cost ratios; yet you recognize these as features of the process described in this unit. So what's so new and different about systems analysis?

The term is ill-chosen, ambiguous, and carelessly bantered about, so frequently, in fact, that it could hardly be ignored here. Confusion is added by frequent use of the terms *systems engineering* and *systems approach* in a similar vein, with similar disagreement among authors. Obviously these terms are more stylish than meaningful.

SELF QUIZ

1. List at least five kinds of things that are specified in the specification phase of the design process.

2. List the five phases of the design process. Also list the additional six events that were mentioned as a part of the full design cycle.

Design Process Phases

Subsequent Events in the Design Cycle

3. What is the function of the implementation events in the design cycle?

4. What is the function of the follow-up events in the design cycle?

5. What is the function of the reactivation events in the design cycle?

1. Some of the things specified during the specification phase of the design process are:

- Dimensions and special relationships between elements of the solution, especially as these can be shown in drawings
- The appearance and actual operation of the design, especially as these can be shown in a mock-up or a working model
- Precise estimates of performance in relation to the criteria and restrictions imposed on the engineer, especially as these can be presented in a report
- Types of materials to be used and their properties
- Tolerances
- Operating conditions such as temperatures
- Methods of assembly, fastening, and so on

2. Design Process Phases

- Formulation of the problem
- Analysis
- Search
- Evaluation and decision
- Specification

Subsequent Design Cycle Events

- Facilitation of acceptance of the solution
- Supervision of installation
- Supervision of initial use
- Surveillance of use
- Evaluation
- Decision to redesign

3. Since technical or economic superiority of a proposal is no guarantee that it will be adopted over other solutions, and because engineering is generally a staff, not a line function, the implementation phase requires the continued effort of the designer to make sure that staff and line approval is forthcoming from the appropriate people, that the project is properly installed and put into use in a way that does not create any problems or prevent the elimination of bugs, and that actual use demonstrates the practical usability of the design as executed.

4. The chief function of follow-up in the design cycle is to help educate the designer so that future designs will be improvements over previous designs.

5. Since no design is ever final, the design cycle turns back on itself and is a completed loop when the point is reached when it would be profitable to find a new solution and the decision is made to redesign. Reactivation of the design process enables designers to take advantage of improved methods, increased knowledge, changed conditions, and many other factors that accumulate after a design has been implemented and utilized for a period of time.

ISBN 0-471-01702-7